Contents

About the Author

Rev. Dr. Dianne D. Glave is an independent scholar whose expertise is black environmental history and black eco-theology. Her latest book is *Black Eco-theology Through History* (2026). Dr. Glave has also published *Rooted in the Earth: Reclaiming the African American Environmental Heritage* (2010) and co-edited *"To Love the Wind and the Rain": African Americans and Environmental History* (2005). She has taught at Carnegie Mellon University, Morehouse College, Tulane University, and Pittsburgh Theological Seminary.

Preface

The socio-economic-political chaos of the world continues to worsen the lives of marginalized people. As such, we must amplify their voices continuing to advocate for them so that those on the margins move closer to the center to equity. Many recognize that climate change triggers extreme heat, fires, and floods with a more severe impact on impoverished people. Rising epidemics, pandemics, and endemics including Covid-19, monkeypox, and polio impact BIPOC (Black Indigenous Persons of Color) more severely. Underlying these environmental and healthcare issues are economic uncertainties worsening access to housing, lowering employment, and increasing inflation. This circle widens with declining and debilitating mental health issues that feed global, national, and local political instability.

Routledge has helped to respond to the chaos through my book *Black Eco-Theology Through History*. Though I am excited about the publication of my new book, I merely state the injustice through historical and theological lenses. Justice remains to be served moving from words to action for BIPOC and the environment. My book leans into the past to begin to tease out a black eco-theology for our consideration and action today.

Past and present, surely some turned to God when the river flooded its banks, spilling and rising into homes of blacks. Past and present, marginalized people including some black children suffer from higher rates of asthma because of increased exposure to air pollution. These environmental impacts remain especially for those at the intersectionality of poverty, race and ethnicity, along with gender, child neglect and abuse, and ageism.

Yet God, whether in historical or contemporary recountings, has been the beacon for many, including those further ostracized in the chaos. Luminous and elegant songs, brave manifestos, heart-felt speeches, intimate poetry, and more tell the stories of black eco-theology. Hope in God eclipsed the injustices in environmental narratives of blacks. Many survived, and further prospered like runaway slave Harriet Tubman who traveled the woods and swamps to freedom. Nat Turner enslaved and an insurrectionist, contextualized rebellion through nature. Black eco-theology reframes and complements their celebrated stories in traditional biographies based in social history.

The perspectives are of people of African descent through religion and nature. They understood nature through God's meaning from personal moments with the divine to prophecy for the community. Blacks engaged in God-Talk, a sacred yet casual conversation with God, a reflection of spiritual and religious beliefs. Through this talk, an omniscient God continues to allow people of African descent to understand in small ways His/Her work in creation, while following God's lead and experiencing life-giving nature.[1]

Black eco-theology also explores how people of African descent depend on God and scripture to mentally, spiritually, and emotionally survive experiences like enslavement set against nature. Blacks lived in nature, while also interpreting the greater meaning of the cosmos including nature. Such a practical theology helped people of African descent make sense of racist violence offset by engaging in the natural world from spiritual perspectives.

In addition, black eco-theology is self-empowering, "a religious approach that emphasizes the interconnections between human spirituality and liberation, environmental awareness and activism, and the special role of the Black population in these processes." This theology interprets nature through Africanisms, Christianity, and activism.[2]

Under the umbrella of black eco-theology is African eco-spirituality, black creation care, black environmental liberation theology (BELT), and wilderness experience theology. African eco-spirituality are complex religious experiences. What follows are generalizations with differences among countries and ethnic groups. African communities depend on strong accountability and selflessness on behalf of the community with individual needs less of a priority. Some Africans treat the earth with respect based on the emphasis on ancestors and agriculture. For example, in Malawi, Africans have built a rain shrine in the Lilongwe forest in Kalembo township. Inside the structure two pieces of black cloth placed above the head and stretched to the walls represent the four corners of the earth. This representation of the earth elevates the importance of nature.[3]

Finally in the natural world, spirituality, ancestors, and daily life remain interconnected, which is very different from Western perspectives with the sharp dichotomy between the secular and the divine. Essentially, religion is here, and nature is over there. Generally, this is not so for African spirituality and African eco-spirituality.[4]

Embracing theory and politics, broadly creation care typically adds action to theology on behalf of the planet. A black proto-creation care reveres nature through the lenses of religion and spirituality. More narrowly, creation care from African perspectives emphasized people and was a precursor to modern creation care launched by whites in the 1990s.[5]

A white proto-creationism began with Augustine through the lens of white male privilege though he was of African descent living from 354 to 430. His extensive written theological record continues to influence Western perspectives of white male patriarchy from the top down and is still steeped in privilege and power. In contrast to Augustine, Sojourner Truth, black, illiterate, and enslaved is known for one powerful narrative and the renowned "Arn't I a Woman?" speech transcribed and published by others. She too wrote deeply and theologically about God's creation.[6]

Black Environmental Liberation Theology (BELT) is advocacy for equity in post-colonized places and with resources for persons of color; whites have long exploited natural resources belonging to others to the detriment of developing nations including those that comprise Africa. BELT is a strand of black eco-theology that defies white oppression originating in a proto-environmental justice in Africa and the Americas. Through such injustice, whites carved out landscapes fractured by racism exploiting nature and people in Africa and later in agricultural fields in the Americas. Whites used the forced labor of blacks to carve up the natural landscape for the monoculture of tobacco, cotton, rice, and indigo. Later in the early twentieth century, whites in part transformed the meaning of nature violently lynching blacks from trees on the landscape. A tree was no longer a tree, but an object that represented violence on a landscape defined and fractured by whites.[7]

Even with such abuse and brutality in nature, blacks in the US held to their beliefs of the just God, the just Creator. A proto-BELT linked to the injustice of enslavement began with the insurrectionist Nat Turner prophetic visions in the context of nature. Starting in the late twentieth century with BELT, blacks developed an expanded environmental consciousness, systematic responses to environmental racism, and collectively a sense of justice acting to protect marginalized people living with disproportionate exposure to environmental hazards like water and air pollution.

Delores S. Williams describes the Hebrew wilderness experience theology in *Sisters in the Wilderness: The Challenge of Womanist God-Talk* (1993). She argues that the theology for some blacks "is a symbolic term used to represent a near-destruction situation in which God gives personal direction to the believer and thereby helps her make a way out of what she thought was no way." This black eco-theology is based on the harrowing biblical narrative of Hagar forced into the wilderness with her son Ishmael. Sarai, a Hebrew matriarch, married to Abram, and unable to have children, gave enslaved Hagar to Abram. After Hagar gave birth to the child Ishmael, Sarai became jealous. Sarai persuaded Abram to send Hagar and Ishmael into the wilderness, more specifically the desert, to die.

According to Williams, "As the result of these hard-time experiences and the encounters with God, Hagar and many African-American women manifested a risk-taking faith." Wilderness experience theology, based in womanist theology reflects on Hagar's experiences that embraces the needs of women, men, children, all part of the community.[8]

Some questions to consider concerning black eco-theology: how did blacks interpret and treat nature from theological perspectives? how have blacks perceived God and Her/His role in nature? how have blacks overcome eco-theological colonialism? how was a biblically-based care for the environment interpreted and practiced by blacks? how did whites dominating black bodies and labor of blacks on racialized landscapes impact the theology and history of blacks? and how did blacks reframe eco-theology and historical narratives of nature making it their own? *Black Eco-Theology Through History* will answer these and many other questions.[9]

The secondary sources are many and critical to the foundation of *Black Eco-Theology Through History* from the perspectives of black eco-theology and people of African descent. Kapya J. Kaoma, Patricia Ngwena, Wangari Maathai, James H. Cone, Marguerite L. Spencer, and Melanie L. Harris are pioneers of black eco-theology. Some call black eco-theology by other names including environmental justice and eco-womanism. My book expands on their important scholarship.

Highlighting Cone's "Whose Earth is it Anyway?" and Spencer's "Environmental Racism and Black Theology: James H. Cone Instructs Us on Whiteness," both conceive of a theology of environmental justice for people of color that continues to be sharpened and nuanced, as the literature to grow. Cone fiercely initiates a theological debate arguing the relationship between justice and eco-theology is central and critical: dismantling racism by whites is inextricably part of saving the planet from destruction. With whites limiting their roles or contributing to environmental racism, BIPOC are adversely impacted. Spencer draws from Cones' essay, reinforcing that environmental justice is the responsibility of whites. She contends that whites must "die to their whiteness," much like Jesus dying on the cross for humankind's sin to dismantle environmental racism. Turning to the New Testament, Spencer echoes the theology of one's old sinful self-dying and becoming new in relationship with Christ to enact environmental justice.[10]

Cone and Spencer knew something was missing. And like me, they may have died spiritually to their old selves made new through a cosmic black eco-theology. I continue to grow into my identity in God, spiritual renewal, and the love for my people of African descent through the heartbeat and a pulse of a black eco-theology. From invisible to visible, I invite the readers to visualize an African in the dampness of the glade seeking God.

Notes

1 Christena Cleveland, *God is a Black Woman* (New York: HarperCollins, 2022); Anthony B. Pinn, *The End of God-Talk: An African-American Humanist Theology* (Oxford: Oxford University Press, 2012); and Delores S. Williams, *Sisters in the Wilderness: The Challenges of Womanist God-talk* (Maryknoll, NY: Orbis Books, 1993).

2 Marta Werbanowska, "'There is Hope in Connecting': Black Ecotheology and the Poetry of Lucille Clift," *ISLE: Interdisciplinary Studies of Literature and Environment*, vol. 26, no. 1 (Winter 2019), 83.

3 James Nathanile Amaze, "The Bimbi Cult in Southern Malawi," Thesis, Degree of Philosophy, School of Oriental and African Studies, 1986. https://eprints.soas.ac.uk/29031/1/10731125.pdf.

4 African eco-theology: Jawanza Eric Clark, *Reclaiming Stolen Earth: An Africana Ecotheology* (Maryknoll, NY: Orbis Books, 2022); Dianne D. Glave, *Rooted in the Earth: Reclaiming the African American Environmental Heritage* (Chicago: Chicago Review Press, 2010), 44; Peter J. Paris, *The Spirituality of African Peoples: The Search for a The Search for a Common Moral Landscape* (Minneapolis: Fortress Press, 1995); Anthony Kanu Ikechukwu, *African Eco-theology: Meanings, Forms, and Expressions* (Independently Published, 2021); Kapya J. Kaoma, *The Creator's Symphony: African Christianity, the Plight of the Earth and the Poor* (South Africa: Cluster Publications, 2015); Wangari Maathai, *Replenishing the Earth: Spiritual Values for Healing Ourselves and the World* (New York: Doubleday, 2010); and Patricia Ngwena, *Eco-Theology from and African Perspective: African Indigenous Life* (Carol Stream, Illinois: Blessed Hope Publishing, 2021).

African American eco-theology: James H. Cone, "Whose Earth it is, Anyway?" *Risks of Faith: The Emergence of a Black Theology of Liberation, 1968-1998* (Boston: Beacon Press, 2000); Dianne D. Glave, "Black Environmental Liberation Theology." *To Love the Wind and the Rain: African Americans and Environmental History*, Dianne Glave and Mark Stoll, eds. (Pittsburgh: University of Pittsburgh Press, 2006); Dianne D. Glave, "Religion: Shouting in the Woods." *Rooted the Earth: Reclaiming the African American Heritage* (Chicago: Lawrence Hill Books, 2010); Melanie L. Harris, *Ecowomanism: African American Women and Earth-Honoring Faiths* (Maryknoll, NY: Orbis Books, 2017); Beth Norcross, "Eye on the Sparrow," *Sojourners Magazine* (August 2012); Eileen M. Smith-Carvos, "Black Churchgoers, Environmental Activism, and the Preservation of Nature in Miami, Florida," *Journal of Ecological Anthropology*, vol. 10 (2006), 33–44; Marguerite L. Spencer, "Environmental Racism and Black Theology: James H. Cone Instructs Us on Whiteness." *University of St. Thomas Law Journal*, vol. 5, no. 1 (Winter 2008), 288–311; Mark, Stoll, "Religion and African American Environmentalism." *To Love the Wind and the Rain: African Americans and Environmental History*, Dianne Glave and

Mark Stoll, eds. (Pittsburgh: University of Pittsburgh Press, 2006); Webanowska, 83–96; and Delores S. Williams, *Sister in the Wilderness: The Challenges of Womanist God Talk* (Maryknoll, NY: Orbis Books, 1993), 108–139.

5 Peter J. Paris, *The Spirituality of African Peoples: The Search for a Common Moral Landscape* (Minneapolis: Fortress Press, 1995).

6 John Doody, Kim Paffenroth, and Mark Smillie, eds. *Augustine and the Environment* (Lanham, MD: Lexington Book, 2016), and Pope Francis, *Encyclical on Climate Change and Inequality: On Care for Our Common Home* (New Port, RI: Melville House, 2015).

Sojourner Truth, "Arn't I a Woman?" Women's Rights Convention in Akron, Ohio, on May 29, 1851 and Sojourner Truth, *Narrative of Sojourner Truth* (Mineola, NY: Dover Publications Inc., 1997).

7 Glave, *To Love the Wind and the Rain*, 189–236.

8 Delores S. Williams, *Sisters in the Wilderness: The Challenge of Womanist God-Talk* (Maryknoll, New York: 1993), 108 and 109.

9 Ta-Nehisi Coates, *Between the World and Me* (New York: Spiegel & Grau, 2015), 12.

10 Spencer, 289–290.

Acknowledgements

In August 2022, Maliya Ellis, a reporter-intern at the Pittsburgh *Post-Gazette*, interviewed me. We met under some trees near Madeline's, a French pastry shop in Regent Square in Pittsburgh. Autumn felt so close but still the sun warmed us; the sunlight streamed around the leaves and branches and touched our hands and arms here and there. Our conversation covered my childhood love of nature to the implications of contemporary climate change on BIPOC to black eco-theology. Maliya inspired me with her openness and curiosity. As a result, I saw my work through an exciting, different, and new lens. For an hour, Maliya became my black eco-theology muse. Through her, I had suddenly found what I had lost: a desire to return to my love and scholarship of blacks and the environment. Our meeting inspired me to recommit to my scholarship.

Before meeting Maliya, a theological and historical spinoff was simmering in my book *Rooted in the Earth: Reclaiming the African American Environmental Heritage* (2010). Two of the chapters focused primarily on religion and other chapters were dotted with religious references. As the book was published, I was on the cusp of graduating with an M.Div. from the Candler School of Theology at Emory University, reinforcing that the next book would combine theology with my background as a historian.

It was not until 2015 that I began to flesh out black eco-theology at a Boston University School of Theology symposium sharing the lectern with Kapya Kaoma, an eco-theology scholar. Later, I contributed an essay titled "Eco-Theology in the African Diaspora" in *The Wiley Blackwell Companion to Religion and Ecology* (2017).

Inspired, I met with Routledge at an American Society for Environmental History Conference. I signed a contract in 2018, which I basically broke because of growing personal responsibilities. In August 2022, I sent the press an email with a plea to work with the press, and they responded within an hour. They reinstated the contract. The book was meant to be.

So many have supported me through this journey. Harold Glave, my father who lived with me, was a source of great emotional support. He has since passed and remained on my mind, a comfort as I finished up the book. Many others include Anthea Butler, Alex Dubov, Derek McKellar, Rebekah Berna, Tom Berna, Elijah Berna, Antolina St. Ill, Wendy Breinig, Jane Fiedler, and Kathleen DeGennaro who provided outlets from writing the book. Going even further back, Mart A. Stewart at Western Washington University, was on my dissertation committee when I worked on my Ph.D. in History at Stony Brook University. Through my struggles as a black woman in the academy, he consistently supported me as I entered the study of the history of blacks and the environment. Finally, the C. C. Mellor Memorial Public Library, along with access to the Pittsburgh Theological Seminary Library, both in Pittsburgh, were critical to completing the book.

The dead, our ancestors, were as important as the living. Insurrectionist Nat Turner and Underground Railroad conductor Harriet Tubman, just to name a few, pulled me back and forth in the past and present communing with them through my research and writing processes. I looked to their words, to tease out a history of black eco-theology. Turner and Tubman are part of a prophetic legacy of people of African descent. We can reflect on the words of both individuals, often turning to and harnessing nature in theological contexts to assuage the impact of racism. Today, my hope is that readers will follow along with me as I interpret their narratives through African spirituality and black experiences in the church and beyond, some of which mirrors my personal life.

1 Creation and Dominion

> Then God said, Let us make humans in our image, according to our likeness, and let them have dominion over the fish of the sea and over the birds of the air and over the cattle and over all the wild animals of the earth and over every creeping thing that creeps upon the earth.
>
> Genesis 1:26 (NRSVUE)

God is black … From the void, S/He created the cosmos including planet earth. According to some black folks, earth was part of the cosmos. Creation narratives with blacks at the center tells of the black God who formed and breathed life into black Adam. S/He gave him broad access to the Garden of Eden. God created black Eve from Adam's rib. God created Adam and Eve in the image of the black God, tasking them to be caretakers of the garden. S/He planted trees and provided food. God created rivers with four tributaries that flowed out of the garden: Pishon to Havilah, Gihon to Cush, Tigris to Assyria, and Euphrates to Assawa. S/He gave Adam and Eve life-giving water to help till the garden. God placed limits on both Eve and Adam: S/HE warned them to stay away from the Tree of the Knowledge of Good and Evil or be punished. The first couple disobeyed those limitations, and God punished them.[1]

Such a black creation theology is about blackness: God, Adam, and Eve align with blackness and pushes against the interpretation of dominion as control by whites. Within these contexts, creation theology is by and for BIPOC' in a world often dominated by narratives of the white God, the white Adam, and the white Eve. Whites have long had power over brown and black bodies in natural and built environments based on dominion, historical racism, and contemporary norms of Christian whiteness. For people of African descent, there is freedom and liberation in taking back and living another narrative about creation.

Balancing injustices against people of African descent, scripture for some including blacks is inerrant with biblical creation for black people. Through inerrancy – The Bible being without human error – God is just, which appears repeatedly from the Old to the New Testaments. Such a

DOI: 10.4324/9780429200700-1

theology of truth and justice is central to the historical experiences of some people of African descent abused during enslavement and with the expectation of justice even if just in the afterlife.

The Bible is also interpreted by people. For example, in creation, all are in the image of God and that includes blacks. The conclusion is rejecting S/He being only white and only for whites. And even with inerrancy, scripture is still subject to interpretation because any reading of scripture is filtered based on gender, ethnicity, age, socio-economic status, and more. Some may deny this, but even simple interpretation is human. For example, the canon, the books included in The Bible, was decided by humans; one book was kept and another discarded in Catholic and Protestant Bibles. Was this process inerrant or interpreted? We also read sacred texts including The Bible, through our preconceived beliefs based in history and culture whether Greco-Roman or African.

Out of such inerrancy and interpretation, black liberation theology informs black creation theology centering the black God, along with black Eve and black Adam in the fall in garden and by extension our world. James H. Cone, says of black liberation theology:

> We have developed a spirituality that plants our feet firmly on this earth, because the God of our faith demands that we bear witness to the humanity of all by refusing to adapt ourselves to the exploitation that the few inflict on the many. We have hammered out a black theology that has helped black churches to recover the authenticity of their faith so that they will not go woolgathering in a nebulous kingdom on high and forget their practical responsibility to live obediently in this world, liberating the poor from the misery of poverty.

Right now, our nebulous kingdom is being destroyed by climate change because of exploitative and ambivalent people. Much of the decisions made about our planet are primarily driven by white men.[2]

Black creation theology and Black liberation theology is also foundational to black environmental liberation theology (BELT). BELT is environmental justice that emerged in the 1980s driven primarily by black church leaders, often male, with black church women doing the critical day to day work. Black environmentalists continue to fight environmental racism including the health impacts of petrochemical plants and garbage dumps. BELT is also rooted in the past: slave insurrectionists like Nat Turner and Harriet Tubman based their connection to God and resistance in part on references to spirituality, scripture, and natural places. Such a theology of black creation informs the activism of BELT, resistance steeped in belief in God, piety, and interpretation of scripture by blacks.

Continuing in the vein of black creation theology and BELT, God is black and just. Christena Cleveland says in *God is a Black Woman* (2022), a journey embracing the Black Madonna and Black Supreme Being,

> She can relate to our experiences of hopelessness because She too has been stolen … As someone who has endured the pain of personal loss and family loss, the Sacred Black feminine empathizes with our pain and stands with us.

God is Black – the axis at the center of the cosmos including the Garden of Eden – so therefore Madonna is Black, the mother of Jesus, so therefore Jesus is Black.[3]

Suffragist Sojourner Truth was in the image of the black God … Race thinker W. E. B. DuBois was in the image of the black God. What if Henry McNeal Turner, the twelfth bishop of the African Methodist Episcopal church who reflected powerfully and theologically that God is black, re-imagined the creation narrative as diverse and in turn centered on justice for marginalized people? If people lived into broadening creation theology and narratives, the world could be more equitable, different, better, and livable.

Genesis 1:27 says, "So God created humans in his image, in the image of God he created them; male and female he created them." Consider the many avenues to interpret this scripture. Whites in power continue to drive and control the creation narrative that long assumed God was only white. The Imago Dei of a black creation narrative – add Asians, First Nation, and Latina/os – begins with the God who is all ethnicities with each of us in Her/His image. Truth, DuBois, McNeal, adding billions more blacks through the millennia are in the image of God. Such theology is black, brown, and white. Vusamazulu Credo Mutwa, who is black and South African, describes Imago Dei in action: "Some say God created Man in His image, so that man should look upon Him as his Father, attend to worldly things and do good deeds. God's very actions prompts us to act to do good deeds."[4]

BELT is evidence of these good deeds. It is cosmic, it is a profound historical and modern re-imagining of the black God creating the sky, water, and creatures in the Garden of Eden; it is a real or imagined place that moved from an equitable Garden of Eden to a present-day faltering broken garden that is our world. The inequity severely impacts BIPOC including those of African descent. The fall of the Garden of Eden is today's climate change, modern creation continuing to be sullied by sin that impacts the planet and people. Can conceptualization of and by extension action by followers of black Imago Dei shore up our damaged world?

Environmental racism worsens climate change for BIPOC and developing nations, which can be corrected through a broader conceptualization of a God of limitless ethnicities. Striving towards a whole or reconstituted modern Eden, everyone has equal access to water … everyone breathes clean air … everyone has easy access to leisure and the amenities of places like Glacier National Park. The God who created the cosmos is still the ruler, maker, and shaker of our declining Garden of Eden, populated by many ethnic groups including black people without equal access to green spaces, who continue to hope for environmental equality. This hope is based on a God for all of us, a return to some semblance of the original Eden, and an opportunity to dramatically reverse climate change.

Reflecting on the many perspectives of Imago Dei, those created by God all have origins on the continent of Africa because it is the cradle of humankind, the origin of humanity. Therefore, before the biblical Adam and Eve, the pair later co-opted as white by whites, there were African creation narratives generally with a Black Supreme God, along with lesser black deities, elders, ancestors, and human worshippers. These narratives may go back as far as 300,000 years to the original homosapiens traced back to Africa.

African cosmology is foundational to dominion and creation stories. Therefore, one must go back to Africa. Through oral traditions, African griots or storytellers shared the myths with many Africans. African myths, including creation myths, in oral traditions, spread as people migrated from Africa – again the location of the origin of humanity – to other parts of the world. The African narratives were later committed to writing beginning in the late 19th century making the narratives more broadly available to the world.

Whether African indigenous beliefs, African indigenous religions, or African Traditional Religion (ATR), Africa's powerful creation narratives are based on myths explaining the natural world interpreting the supernatural. The myths, including creation narratives, explain impossible extraordinary phenomena so ordinary people can comprehend the cosmos. The same was true for biblical narratives. Stephen Belcher amplifies creation as myth in *African Myths of Origin* (2005) "A story that explains origins and root-causes of things is called a myth, and the word has acquired a tremendous range of associations ranging through the religious, spiritual and psychological domains." African creation stories focus on the origin and creation of humanity; it is the relationship with the supreme being and the disruption of that relationship because of human transgressions.[5]

African and biblical creation myths differ and are similar. One story that differs is about the Pygmies from the forests of Cameroon to Rwanda. Khvum, the creator, was alone and wanted to create people for company. He attached a harness between his canoe and a crocodile and headed out into the water. From his canoe, Khvum stopped to throw nuts towards the land. The nuts became people back on shore. When Khvum returned to shore, he happily lived among the people. What's different compared to scripture? In Genesis, God began by creating just two people. The numbers grew when Eve gave birth to her descendants.[6]

In another myth that differs, Eshu – a trickster of the Yorùbá, a West African ethnic group in Togo, Nigeria, and Benin – assisted The Father in creating humans. Eshu had no responsibilities so committed all of his time to the work. He "stayed and helped the Father as he carved the various beings that would fill the world." The Father rewarded Eshu for such diligence creating humans. The Father directed Eshu to the crossroads, between the supernatural and the human worlds, so that everyone coming and going would give Eshu gifts. Looking to scripture, God worked alone, without any assistance, from the creation of the sky to water to animals to humans. What differs compared to the Genesis account is God did not have an assistant.[7]

As noted earlier, African narratives long preceded creation accounts from other ancient civilizations. Western scholars have drawn similarities between creation narratives in biblical and other ancient accounts. *In Quest of the Historical Adam: A Biblical and Scientific Exploration* (2021), William Lane Craig, attributes the origins of the biblical narratives to the Ancient Near East (ANE), saying,

> OT scholars have been aware for centuries of the resemblance of the primeval narratives of Gen 1-11 to ancient Mesopotamian myths on the basis of the Babylonian priest Berossus' account of Mesopotamian religion for Greek readers of the third century BC.

African creation narratives preceded Berossus' interpretation in *History of Babylon* (2019) ancient city in Mesopotamia, along with Greek and Hebrew biblical narratives of creation. Unfortunately, out of biases, really racism, Craig and the OT scholars failed to go back even further to ancient African narratives that were definitive and critical in all comparisons of creation narratives.[8]

Esther Bamitale Olumide argues much the same in "A Comparative Analysis of Genesis Creation Story and Yorùbá Myth of Creation," saying,

> However, it is not difficult to notice the striking similarities between the two creation narratives such as the motifs behind the creation of heaven and earth, separation of heaven from earth, the molding of human from the dust of the earth and the creation of the world in six days and rest on the seventh day.

Olumide adds that the Yorùbá most closely parallels the Genesis narratives when compared to these "other theories of creation" with one God and creator, creation of the world out of the void, relationship with God, and severed relationship between God and humanity. In Yorùbá yore, "Olódùmarè created the universe out of chaos (Genesis 1:2) and that Ọbàtálá was assigned to mold the physical human body from the dust of the earth and Olódùmarè gave life to the physical body (Genesis 2:7)."[9]

Again, Genesis 1 and 2 share similarities with African creation narratives, arguing the latter as the earliest narratives known to humankind. Drawing from The Bible, there or several overtures or sections of the narrative: creation of the earth; lightness and darkness; a dome separating the waters – rain and oceans; flora and fauna; night and day; humans made in God's image; the Garden of Eden; human dominion over it all; sin against God by humans; and God turned humans out of the garden.

In *African Folk Tales: Indaba, My Children* (1964), Vusamazulu Credo Mutwa collected oral narratives from the Bantu, a people who live in Central to Southeast to South Africa. Mutwa honored his family on his mother's side and his people documenting sacred African myths, many focusing on creation. One narrative, in the form of a poem, is titled "The Sacred Story of the Tree of Life," surely a precursor to the Tree of Life narrative in The Bible. Genesis 1:6-8 (NRSVUE) mirrors this African narrative:

> No stars were there—no sun,
> Neither moon nor earth—
> Nothing existed but darkness itself—
> A darkness everywhere.
> Nothing existed but nothingness.

Genesis 1:1 describes a similar void, a nothingness, where "the earth was complete chaos, and darkness covered the face of the deep."[10]

The Genesis 1 and 2 narrative echoes another African narrative outlining various aspects of heaven and earth including the creation of the two places. In the 14th and 15th century Kingdom of Bunyoro, near Lake Victoria in what is now Uganda, Ruhanga, the creator along with his brother Nkya discussed creation. Nyka challenged Ruhanga pointing out that the creator had not done all the things that had been promised. Ruhanga responded by making many things: the sun, clouds, moon, sunlight and darkness, and rain. Both brothers lived in their Eden, and "earth and heaven were close in those days; heaven was held up by a tree, a pole and iron bar." In comparison, Genesis 1:1-2 says, "When God began to create the heavens and the earth, the earth was complete chaos, and darkness covered the face of the deep, while a wind from God swept over the face of the waters." In biblical creation, God created many things similar to the creator in the Bunyoro narrative including darkness.[11]

The creation narrative of Kingdom of Bunyoro includes the creation of the earth. The same is true of the people of the Kuba Kingdom in what is now the eastern Republic of Congo, adding the creation of animals. The great spirit "Mboom vomited up the crocodile, the goat, the egret, the minnow, the leopard, the crested eagle (associated with the royal clan) and men, along with other animals. These beings in turn vomited other beings: the crocodile produced all the snakes and reptiles, the goat vomited all the horned animals, the minnow all the fish, and so on… Mboom then visited the establishments of the different beings and ordained their way of life." Making parallels and though less specific, Genesis 1:24-25 tells us that God created wild animals, creeping things, and cattle.[12]

Again, Africans and biblical creations narratives differed and were similar. Dominion is very specific to the Hebrew biblical narratives later embraced and amplified by modern western civilizations. This idea and practice of dominion is not explicit in African creation narratives.

Focusing for now on scripture, biblical creation narratives were written somewhere from the 6th to the 5th centuries BCE when the Babylonians conquered and subjugated the Jews. The Babylonians exiled the Jews from the kingdom of Israel, and later the Persians conquered Babylonians with a kingdom that include rule over the Jews. The Jews learned about dominion – control by one group over another group – through the Babylonian and Persian empires colonizing the Jews as unwilling subjects. Therefore, the Hebrew writers of Genesis knew firsthand about dominion or control, which became central to the biblical creation narratives.

Looking to the Hebrew – the primary language of the Hebrews, Jews, or Israelites, the second being Aramaic – the meaning of dominion is similar in the context of captivity by the Babylonians and rule by the Persians. Radah or dominion in the Old Testament Hebrew, historically meant to subdue an enemy. Said another way, an individual or group wielded power over others. Jews in the ancient Middle Eastern world understood dominion meant to subdue hostile individuals or groups. The Babylonians and the Persians subdued the Jews. Dominion was woven into biblical creation narratives based out of lived experiences by the Jews.

From the medieval period, the 5th to 15th centuries and beyond, the meaning of dominion expanded moving from the micro of local enemies to the macro of the world: many interpreted dominion to broadly mean control and exploitation of people and resources. Christians ignored and negated the caveat that God expected dominion to be wedded with good stewardship of creation. From the late 20th century, modern theologians formally began to use dominion language.[13]

Dominion is central to the three biblical creation narratives: the Priestly, Yahwist, and Noahic. The Priestly and the Yahwists narratives are arguably different perspectives on the first creation in scripture, and yet the same core story. The Noahic is the second creation narrative. In the

Priestly and Noahic versions, God gave humanity dominion or control of the earth.

Genesis 1, the Priestly version, is a positive take on the human origin experience; Genesis 2, the Yahwist and darker version, emphasizes the fall of humankind; and Genesis 6–9, the Noahic, depicts the second creation after the destruction of the earth by flood. Taken together, these narratives are foundational to the modern interpretation of the dominion of nature including people. Today, such a foundation is reinforced around the world as white theology, history, and experience permeates societies directly and indirectly defining the control and treatment of resources like water in the ocean, timber in the forests, and oil underground.

The Genesis 1 priestly version unspooled as a mythic, gentler, and sublime idyll, with God lovingly making creation. Wind, water, and light were evoked. The water of the sky and sea were separated by a dome. Water was drawn back to reveal land, where God plants seeds. The light of the sun, along with that of the stars, separated day and night. Living creatures were then introduced. Finally, God created humans to have dominion over living things. The narrative is both spectacular and soothing.

In Genesis 1, dominion is gentle. God was the landlord renting to the garden, the world to Adam and Eve. God let Adam and Eve have dominion, essentially access to the agricultural riches of orchards and farmlands reflecting the actual lives of the scribes of Genesis 1 living in ancient agricultural societies under Babylonian and Persian rule. Genesis 1:26 says, "Let them have dominion over the fish of the sea and over the birds of the air and over the cattle and over all the wild animals of the earth." Again, the interpretation of dominion of let is key.[14]

Conversely, the Yahwist version was more dire and less sublime. The Garden was a rental gone wrong. Adam and Eve were tenants tasked with upkeep, with the possibility of eviction being imminent. Keeping to this tenancy metaphor, yes, Adam and Eve's situation was not permanent. God told them to stay away from the Tree of Knowledge of Good and Evil, which the couple ignored eating fruit from the tree. Ultimately, God banished Adam and Eve from the Garden of Eden because the first couple sinned against God including lying. The ejection continued the metaphor of a landlord evicting a renter from a rental. What landlord wanted dishonest tenants?

God revoked the temporary rights to a sublime life in the garden. S/He condemned Adam and Eve's to struggle tilling the earth. Said another way, the couple's legacy of sin became painstaking and brutal harvesting to simply produce a kernel of corn or grain of wheat. Interpreting this biblical legacy for today, we face climate change out of the first sin, where an abuse of the planet struggles to sustain humans. People are in an environmental struggle worsened for marginalized including developing nations and BIPOC. Exposure to pollution or forced migration because of floods is higher for marginalized countries and ethnic groups compared to developed nations and whites.

Adding the Noahic destruction and creation to the Genesis narratives, the theme of dominion continued. In Genesis 6-9, God sent a flood that covered the world punishing humanity for their unrepentant sins, a recurring theme in the Old Testament. God spared Noah and his family, along with all sorts of flora and fauna to repopulate the earth after the flood. Essentially, God once again re-created the world after Genesis 1 and 2 and the destruction by flood. When a dove was sent out and returned, it was assumed there was no place for it to land on dry land with water still covering the earth. Eventually, a dove did not return indicating it found dry land. After Noah and his family disembarked from the ark, God told them to "bring out with you every living thing that is out of all flesh–birds and every creeping thing that creeps on the earth."[15]

Grateful his family was saved, and the earth was renewed or recreated, Noah gave a burnt offering and God decreed as part of the Noahic covenant in Genesis 6:7, "And you, be fruitful and multiply, abound on the earth and have dominion over it." God blessed Noah and his sons and said to them, "Be fruitful and multiply and fill the earth. The fear and dread of you shall rest on every animal of the earth and on every bird of the air, on everything that creeps on the ground and on all the fish of the sea; into your hand they are delivered. And you, be fruitful and multiply, abound on the earth and have dominion over it."[16]

As a reprieve by God, flooding returned the earth to times before Adam and Eve began to populate the earth. Noah and his family's narrative is a second creation, reaffirming dominion in The Bible. God used Noah to redeem a world gone wrong, a sinful world, through Noah's obedience to God by building and launching the ark. Today, the Noahic covenant signifies a second chance for innocent marginalized people including those of African descent, loosely making a comparison to Noah and his family. In the tradition of justice, the guilty are accountable, based in the metaphor of the sinners who died in the flood in Noah's time.

The Genesis creation narratives – the Garden of Eden and the aftermath of the Noahic flood – are the underpinnings for contemporary dominion and creation. Beyond the ancient biblical creation narratives, whites have long controlled the analysis, interpretations and actions of dominion and creation. Whether European, Anglo, European American, or white, they primarily benefit from dominion – exploitation and destruction of creation – upkeep and rehabilitation of the earth. African and African American perspectives temper those of the dominant culture.

Contemporary dominion of nature has profoundly transformed our planet. Climate change is driven by the legacy of dominion of nature practiced by Christians. That includes BIPOC. Still whites continue to grasp at and control limited resources and desirable and healthy green places. In addition, the planet is at a crossroads and a large part of the decline is based in explicit and implicit white Christian mores that devalues people and the planet, treating both as expendable and endless resources.

As early as 1967, Lynn White, Jr., a professor of medieval history, sounded the alarm about the history of Christian dominion that continues into the present. In "The Historical Roots of Ecological Crisis," White says,

> We are superior to nature, contemptuous of it, willing to use it for our slightest whim ... The whole concept of the sacred grove is alien to Christianity and to the ethos of the West. For nearly 2 millennia Christian missionaries have been chopping down sacred groves, which are idolatrous because they assume spirit in nature.[17]

White traces the impact of whites on the environment from the Middle Ages to the Industrial Revolution, which extends to the present. In the early medieval period, agriculture had less of an impact on the planet: perhaps unwittingly the peasants practiced sustainability limiting to working a share of the land just to feed their families. He says, "Early plows, drawn by two oxen, did not normally turn the sod but merely scratched it" and that "formerly man had been part of nature; now he was the exploiter of nature." White adds,

> Peasants pooled their oxen to form large plow-teams, originally receiving (it would appear) plowed strips in proportion to their contribution. Thus, distribution of land was based no longer on the needs of a family but, rather, on the capacity of a power machine to till the earth.[18]

By the mid-19th century whites practiced industrialized agriculture, expanding on the agricultural practices of medieval peasants. Whites of the era, turned to cotton gins, first invented in India. The machine mechanized agriculture and increased production that separated cotton from the seeds. The number of people working by hand dropped from many to a few. Such technology ramped up cotton production, razed wooded areas, and replaced fields with cotton production. The cycle exhausted the soil of nutrients, and in some cases left fields fallow and exposed to soil erosion. White asks, "Is it coincidence that modern technology, with its ruthlessness toward nature, has so largely been produced by descendants of these peasants of northern Europe?" Those peasants were and their descendants are white.[19]

In the 21st century, industrialized agriculture continues to accelerate environmental degradation as companies like Dole and Del Monte put food on grocery store shelves using technology to plant vast amounts of engineered cash crops like corn by deeply plowing the soil, diminishing soil fertility, and using chemical fertilizers to return fertility. The result is soil erosion and more chemicals in the soil and water.[20]

Throughout White's essay, he references Europeans destroying nature. He does not specifically name whites, including Euro-Americans, but they are primarily responsible for the history that got us to climate change. Such dominion of nature is framed and reinforced by whites though BIPOC are part of the broader culture engaging in the same views of practices. Blacks have drawn on a nuanced mix of influences including Africa, The Bible, and dominion. Yet Whites, Christians, and neo-Christians continue to be at the forefront of exploiting and destroying the planet and brutally harnessing the labor of marginalized people. Even today, those who practice dominion, benefitting or suffering, have forgotten that God mandated people care for the planet.[21]

Generally, many live in nature as the exploiter versus the exploited or neglected. The tension does no good for the planet. Whites exploit and neglect BIPOC including marginalized people in developing nations. The exploitation is institutionalized, deeply ingrained in the world's actions including the US. Whites, with the most power and wealth, have been most harmful to the environment and people. Institutionalized oppression is to the detriment of BIPOC and the environment. Adding another layer, oppressed groups and individuals oppress others and nature. The collective, whether white or BIPOC are accelerating climate change on the planet, based in white Christian and neo-Christian behavior and dominion of vulnerable people and resources.

Whites have woven dominion as control, utilization, and destruction of black bodies in the environment. The former have enriched themselves wielding the labor of marginalized workers in factories and large-scale farming. These laborers have been used to transform and worse pollute and transform nature enriching some whites.[22]

Many have wielded systemic use of dominion against others, essentially the oppression of enslaved people of African descent against the backdrop and as a means of harnessing nature. Dominion through financial gain was set in environments of Middle Passage and enslavement. The Portuguese and the Dutch followed by the French and British, captured and enslaved Africans, human commodities, delivering the enslaved Americas and the Caribbean to cultivate cash crops in the built agricultural environments. The use of blacks as a group by whites for financial gain has transformed the US.

Contemporary films of the Middle Passage and enslavement follow the fictionalized and historical. The cinema reinforces images of abused black bodies and pain against natural and built environments of the holds of slave ships, fields of cotton, and even slave insurrections. "Amistad" and "12 Years a Slave" create today's spectacle and entertainment in never-ending loops of black bodies suffering the vermin on ships and the backbreaking labor in cotton fields.

For financial gain, enslavers used Christian dominion to indoctrinate and manipulate Africans by twisting scripture to do this unpaid work in these built environments. A portion of Ephesians 6:5-8, specifically verse 5 says, "Slaves, obey your earthly masters." Whites including slaveholders interpreted these words as a spiritual rod and manipulation to brutalize and to control the enslaved. Reading deeper into the chapter, whites left out the context of being slaves of Christ including obeying Christ and doing the will of God. Slaveholders treated the scripture literally rather than metaphorically: being a slave is positive in a spiritual context by deepening one's relationship with Christ. Slaveholders exploited black bodies reinforcing exploitation through scripture to force the enslaved cultivate rice, cotton, indigo, and tobacco for no wages.[23]

Toni Morrison's *Beloved* (2004), Billie Holiday's in "Strange Fruit" (1937) and Ta-Nehisi Coates' *Between the World and Me* use the language of black bodies. Though the three do not mention dominion of black bodies, there are connections. In the novel, Baby Suggs, a runaway slave and black preacher is direct with her congregations who probably escaped enslavement in the south too:

> Yonder they do not love your flesh. They despise it. They don't love your eyes; they'd just as soon pick em out. No more do they love the skin on your back. Yonder they flay it. And O my people they do not love your hands. Those they only use, tie, bind, chop off and leave empty.

The reference has a double meaning: the first is the physical destruction of those hands and the second is the work in fields binding sheafs of tobacco, chopping cotton, and leaving the crop leaves the enslaved hands to be sold by whites leaving the slaves with nothing.[24]

As was true for Morrison, the subtext is dominion of black bodies by whites. The same is true for Billie Holiday's recording of "Strange Fruit" in 1937, a resistance song against lynching, which escalated in the early 20th century. Traditionally the strange fruit, a metaphor for lynching blacks from trees, can also be extended for today as commodification and brutalization of black and brown bodies for profit, following the pattern of dominion.[25]

In the horrific tradition of dominion and commodification, Coates traces some of that same violent history of the exploitation of black bodies, built environments, and financial gain by whites:

> At the onset of the Civil War, our stolen bodies were worth four billion dollars, more than all of American industry, all of American railroads, workshops, and factories combined, and the prime product rendered by

> our stolen bodies–cotton–was America's primary export. The richest men in America lived in the Mississippi River Valley, and they made their riches off our stolen bodies. Our bodies were held in bondage by the early presidents. Our bodies were traded from the White House by James K. Polk. Our bodies built the Capitol and the National Mall.[26]

Whites still practice dominion of people and the environment. Whites treat neo-enslaved and post-colonized landscapes and bodies with disregard. Such racism, specifically environmental racism, is echoed in the aftermath of hurricanes in Puerto Rico, the Bahamas, and New Orleans: people of African descent experience the environmental brunt of the aftermath of the devastation resulting from climate change.

Yet African Americans have been more than brutalized bodies, more than victims in Christian and Christianized worlds. They are victorious, taking back their humanity when the world treats them like animals in built environments and nature. Looking once again to *Beloved*, Toni Morrison's reference to flesh is synonymous with black bodies and is in a spiritual context. Baby Suggs' message or sermon blends whites abusing black flesh with a hope for the members of her black Christian community. Suggs says, "'Here,' she said, 'in this here place, we flesh; flesh that weeps, laughs; flesh that dances on bare feet in grass. Love it. Love it hard'." With all the emotions embodied in the flesh, place is many things for her congregation: the wooded area where her congregation worships; the state of Ohio to where they have escaped the brutalities of enslavement in the South; and the yearning for that place called heaven. The woods, Ohio, and heaven are theirs, and the South is a nightmare landscape of the past.[27]

Creation and dominion can be traced in ancient history and the contemporary history. The Genesis and African creation differ and have much in common, with roots in Africa. Further along the timeline, whites have wielded dominion against brown and black bodies, along with nature as exploitation. Such knowledge calls for reversal of dominion by whites for the sake of blacks and the planet. Blacks continue to take back their bodies and nature, further caring for nature in a world where dominion alone centers damaging environmental thinking and practice.

Notes

1 Many versions of The Bible use different terms in Genesis 2:15: NRSVUE – till; NIV – Care; MSG – work; and NKJV – tend/keep.

2 James H. Cone, *For My People: Black Theology and the Black Church* (Maryknoll: Orbis Books, 1984), 192 and Cornel West, *The Cornel West Reader* (New York: Basic Civitas Books, 1999), 393.

3 Christena Cleveland, *God is a Black Woman* (New York: Harper One, 2022), 237.

4 Genesis 1:26a, NRSVUE: Then God said,

> Let us make humans in our image, according to our likeness and 1 John 3:2 (NRSVUE): Beloved, we are God's children now; what we will be has not yet been revealed. What we do know is this: when he is revealed, we will be like him, for we will see him as he is.

Genesis 1:27, NRSVUE and Vusamazulu Credo Mutwa, *African Folk Tales: Indaba, My Children* (New York: Grove Press, 1964), 537.

5 Stephen Belcher, *African Myths of Origin* (New York: Penguin Books, 2005), xiii.

6 Belcher, 18–19.

7 Belcher, 97.

8 William Lane Craig, *In Quest of the Historical Adam: A Biblical and Scientific Explanation* (Grand Rapids: William B. Eerdmans Publishing Company, 2021), 22 and Esther Bamitale Olumide, "A Comparative Analysis of Genesis Creation Story and Yorùbá Myth of Creation," (Kwara State University (Nigeria) ProQuest Dissertations Publishing, 2019), 2.

9 Olumide, 2 and Genesis 2:7, NRSUVE.

10 Genesis 1:6-8, NRSVUE; Vusamazulu Credo Mutwa, *African Folk Tales: Indaba, My Children* (New York: Grove Press, 1964), 159 and Genesis 3:22, NRSVUE; and Genesis 1:1, NRSVUE.

11 Belcher, 171 and Genesis 1:1-2, NRSVUE.

12 Belcher, 215–216.

13 Lynn White, Jr.,"The Historical Roots of Our Ecological Crisis," *Science*, vol. 155 (1967), 1203–1207. https://www.cmu.ca/faculty/gmatties/lynnwhiterootsofcrisis.pdf.

14 Genesis 1:26b, NRSVUE.

15 Genesis 8:17, NRSVUE.

16 Genesis 9:7, NRSVUE and Genesis 9:1-2, 7, NRSVUE.

17 White, 1203–1207.

18 White, 1203–1207.

19 White, 1203–1207.

20 White, 1203–1207.

21 White, 1203–1207.

22 Ta-Nahesi Coates, *Between the World and Me* (New York: Spiegel & Gra, 2015), 35–36.

23 Ephesians 6:5-8 (NRSVUE).

24 Toni Morrison, *Beloved* (New York: Vintage Books, 2004), 103.

25 Billie Holiday, "Strange Fruit," 1939 and Ephesians 6:5-8 (NRSVUE).

26 Coates, 101.

27 Morrison, 104.

2 Creation Care

> The LORD spoke to Moses on Mount Sinai, saying ... You shall not reap the aftergrowth of your harvest or gather the grapes of your unpruned vine: it shall be a year of complete rest for the land.
>
> ~Leviticus 25: 1 and 5

Imagine Ned Cobb, a God-fearing African American sharecropper magically transported from Alabama in the United States from the 20th century. He is suddenly a Jew during 6th century BC antiquity in the southern kingdom of Judah. Through time, cultures, and continents, the real and imagined Cobb was devout. No matter the millennia, imagine him following God's laws for caring for creation based on the book of Leviticus and other passages of the Old Testament. Imagine the agricultural worlds of the lush American South with rainfall and the hot and arid Middle East. Imagine for Cobb, creation including the land was sacred in both places. Imagine that time traveling, Cobb dispossessed and disenfranchised as a black sharecropper, understood the burdens of the widow and the orphan in Judah and Alabama. To the mainstream those with the greatest needs were often invisible like Cobb for much of his real life against the backdrop of the environment.[1]

Perhaps time-traveling and landing in Judah, Cobb might have had different choices. As a Jew, he would have cultivated grape vines for wine and olive trees for oil with some vegetable patches – all terraced on hills. Any produce from the land was hard fought and won on the daunting landscape. Hands still brown – darkened by the desert sun – and etched with soil on his cracked hands and under nails, he might have practiced as a Leviticus 25:5 farmer in Judah. Cobb followed the will of God leaving leaves and vines to return nutrients to and the land fallow to replenish the soil.

White theologians have been foundational and dominant in care of creation theological thought compared to disenfranchised blacks like Cobb. Yet Cobb was a black forerunner of creation care as a practical theology. Whites no longer remain the echo chamber of such theology.

DOI: 10.4324/9780429200700-2

Cobb's proto-creation care theology evolved into a contemporary creation care defined by and for the fulfillment of whites. Creation care is understanding and living into the sacredness of God-given earth's flora and fauna: the birds, the trees, the air, the water. It is difficult to pinpoint the origin of contemporary creation care in the 20th century; yet it continues to be a theology of praxis including buying produce locally, purchasing less goods, and protecting watersheds. White Christians continued to advocate for the planet even before the devastation of early 21st century climate change of Hurricanes Katrina and Sandy. The movement is primarily elite white activists. Seeing creation is in crisis, they respond to God's call to intervene on behalf of nature based on biblical principles. Christians have sinned against God and the planet by continuing to destroy the soil, air, and water. White Christian activists seek to reverse the destruction of nature with limited concerns about the people.

Steven Bauma-Prediger says, "Authentic Christian faith includes care for the earth. Earthkeeping is integral to Christian discipleship." The question is authentic to whom? The answer is primarily white Christians. Generally, environmental justice, a component critical to BIPOC survival in the environment, is often missing from white care of creation theology, dialogue, and practice. As a result, mainstream creation care focuses advocacy primarily on nature and non-human creatures, and not marginalized people. The creation care tent must be expanded beyond privileged whites for equality for BIPOC in nature. A more fully developed interpretation of contemporary creation care by the Christian faithful at a minimum and in tandem advocates for and acts on behalf of both people and the environment.[2]

Whether through scripture, early history, or in the present, God gave people the gift of the earth and commanded that we honor the gift. In this vein, Genesis 2:15 is the foundational creation care scripture: "The Lord God took the man and put him in the garden of Eden to till it and keep it." The Common English Bible uses care rather than keep. "Keep it" or "take care of it," is based on the Hebrew šāmar or care. Šāmar is to watch over, protect, and guard. Again, the scripture is critical yet scarce on details on how to execute taking care of creation.[3]

Whether black or white, creation care is based on scripture. In the ancient Middle Eastern world, through scripture God exhorted that people care for the environment in Leviticus 25:5. Jews cared for the environment living by the Torah, the law of Genesis, Exodus, Leviticus, Numbers and Deuteronomy. Doing so, honored and glorified God.

Out of Jewish religious tradition, God used prophets like Moses, the author of Leviticus, to share the rhema, God's word, with the people of Judah. Leviticus 25:5 is silent on the laborers who managed daily operations or directly farmed acreage. Hearing the Word, God's people were to be obedient and act: leave the land fallow, give the land time to rest, and in turn renew its fertility. The call of the 21st century is caring for people first and the land or nature second, an interpretation and expansion of the rhema of Leviticus 25:5. More expansively and for today, the rhema of Leviticus is for people who are dispossessed blacks like Cobb who are the most vulnerable to climate change and disasters, who need the most care. Moving from dispossessed to having the leisure to care and act on behalf of the environment reverberates today.

As there are many millennia, so too are theological reflections or interpretations of God's creation throughout history to the present. The nascent or proto-creation care theology has long been attributed to white men beginning with antiquity including Augustine of Hippo (354–430) and Francis of Assisi (1181–1226).

Augustine of Hippo was rational in his theology and Francis of Assisi was poetically reverential about God and creation. Creation was a subject, critiqued and loved in early European history. Such distance to a subject, whether analytical or appreciative, was not about action nor justice for the planet that is typical today. Unlike the present-day, Francis and Augustine of medieval times were not facing global cataclysmic climate change. Therefore, creation care was not prevalent and practiced in the contemporary sense and not critical as resources seemed limitless with smaller populations from antiquity to the Middle Ages. Theirs was a predecessor to care of creation.

For many, Augustine remains a revered saint of the Catholic Church, and perhaps the most influential theologian in Western Christianity. His father was a Roman colonist, and his mother was Berber, an ethnic group of North Africa. Still, Augustine self-identified as a Latin speaking Roman, and his theological perspectives, his writings and beliefs was that of a Roman conqueror. His work continued to be part of Western civilization though, foundational in European Christianity, and later continuing into and embedded into 21st century North American religion and culture.

Augustine interpreted nature through duality and hierarchy. Both are based on Greek Platonism and medieval Christianity, today still foundational to Western Christianity. Augustine's interpretation of creation is theologically dualistic, which can be interpreted broadly and theologically to include God versus Satan. For Augustine, in the context of nature, dualism is God versus the created universe. Mark Rochelle Ferraren Renacia says for Augustine, "there is a hierarchy of perfections with the spiritual realities being superior and possessing authority over the material," adding "the result is an insurmountable gap between the creator and the creature." Augustine interprets God as distanced from creation, and in turn humans distanced from creation. And among the creatures of God a higher value is placed upon spiritual and intellectual creatures – that is humans – because they are by nature closer to God. The God/creation hierarchical dualism can be interpreted as God versus people versus creation. God and by extension people "with spiritual souls" are different from nature. Augustine reinforces this hierarchy through God's greatness in connection to nature:

> Therefore you must have created them from nothing, the one great, the other small. For there is nothing that you cannot do. You are good and all that you make must be good, both the great Heaven of Heavens and this little earth. You were, and besides you nothing was. From nothing, then, you created heaven and earth.[4]

The West including Christians continue to practice a platonic and Augustine dualism that divides people into a hierarchy based on ethnicity and socio-economic status. Lower in the ranking are creation and marginalized people, lowest on the hierarchy in Augustine's time and millennia of climate change. A hierarchy, authority, and power worsens climate change and the plight of marginalized people. Dualism and hierarchy are a caste system that begins with God, followed by people, and finally nature. What remains is a caste system within a caste system that begins with white male privilege and power and ends with and disenfranchises women, the elderly, the impoverished, the unhoused, BIPOC, developing nations – all against the backdrop of nature.

Augustine's hierarchy of creation is a systematic theology and is different from St. Francis of Assisi's blissfully emotionally inter-connected to nature. Francis was a Catholic and an Italian mystic–the latter another distinction from Augustine–who practiced a vow of poverty and founded the Franciscan friars, Catholic monks. His poem "Canticle of Brother Sun and Sister Moon," edges closer to contemporary creation care as a proto-creation care, through poetic appreciation without activism. Francis makes himself a lowly conduit between God and creation, praising God whereas Augustine elevates humans. The canticle is a love song to God and nature. Francis says,

> Praise be You my Lord through our Sister,
> Mother Earth
> Who sustains and governs us,
> Producing varied fruits with colored flowers and herbs.

Francis praises God through Mother Earth, Sir Brother Sun, Sister Moon, Brother Wind and Air, Sister Water, and Brother Fire. In 1979, Pope John Paul II declared Francis the patron saint of ecology because of Francis' kinship with nature, looking to the poem as an example. As a saint, Francis has influenced the contemporary world of Christian spirituality and religion.[5]

The care of creation timeline continues into the 21st century with Pope Francis, once the leader of the Catholic Church. Pope Francis went further than early care of creation advocates: his theology is justice for creation. Pope Francis explicitly extended justice to those who are marginalized because climate change has worsened environmental exposures. I wonder if Augustine, who led a privileged life, could have conceived of such justice. Perhaps Francis of Assisi might have as he took a vow of poverty and cared for the poor.

Pope Francis acknowledges that "many intensive forms of environmental exploitation and degradation" has the planet on the brink of climate change exhausting precious natural resources. With environmental injustice and racism, the pope advocates for the disadvantaged: "The disappearance of a culture can be just as serious, or even more serious, than the disappearance of a species of plant or animal."[6]

Unlike St. Augustine and St. Francis – the comparisons are a stretch because of different times and places – Pope Francis encourages action, through dialogue partners and his own interior conversion. He directs the faithful to participate in dialogue partnerships with indigenous communities. Rather than deepening the leading role of whites through their privilege, one must follow the lead of indigenous communities tending to and living directly in creation. Pope Francis says, for indigenous people,

> land is not a commodity but rather a gift from God and from their ancestors who rest there, a sacred space with which they need to interact if they are to maintain their identity and values. When they remain on their land, they themselves care for it best.

As part of this sensitivity and support of indigenous people, the pope calls for a relationship with Jesus Christ that extended to nature, to the planet. Such an "ecological conversion" translates into action by protecting "God's handiwork is essential to a life of virtue; it is too proportional or a secondary aspect of our Christian experience." Pope Francis' words go beyond platitudes encouraging activism and engagement.[7]

St. Francis, St. Augustine, and Pope Francis, all white men – fall short of the challenges in creation experienced by marginalized people including BIPOC. Pope Francis urged action but much of what he did remains in the theological realm without his own hands-on action. Popes lead, the followers are to act. God calls people to act to protect people on planet earth.[8]

Pope Francis' focused on indigenous people; add too people of African descent. Whether indigenous or black, the faithful are to intentionally care for creation, serve and protect marginalized people who suffer the most because of climate change.

Black creation care is one aspect of black eco-theology, going beyond creation care defined and led by whites. Brown and black bodies, brown and black people who have long labored as enslaved and free, enriching privileged nations and individuals are endangered. People are an integral part of creation, part of earth keeping. Yet some historically have not been prioritized especially BIPOC.

African creation care – some might simply call this a way of life – was a precursor to a national movement of creation care beginning in the US during the 1990s. There is no creation care without acknowledging that Africa, the cradle of and the first civilization, was one of the first to place creation at the center with origin stories based in African cosmology. That said, black creation care embraces and rejects white creation care. Esther Bamitale Olumide argues, "It equally aims at assessing the possibility of interpreting creation accounts in the Book of Genesis in the light of the Yorùbá myth of creation."[9]

From the perspective of people of African descent, what is critical whether a follower of Christianity, African spirituality, or both, is that people of African descent are central to the creation story through the lens of being in God's image. God is black to some black people centering blacks even more in stories based on religion and spirituality.

Again, with blacks in mind, scripture can be used to diminish or elevate. During the period of enslavement, black ancestors, including Cobb's people, suffered. In the United States, Christianity has been woven into our culture, even among those who do not profess Christianity. Interpretation of scripture has been manipulative, taken out of the context in different eras. For example, "Slave obey your master," was weaponized by slaveholders to control people of African descent who were forced to work the land for free to the financial benefit of whites. Obedience was different from enslavement in the ancient world, where a former slave could transition into a powerful member of society in antiquity. Not so once enslavement was racially codified in the Americas and the Caribbean. Blacks like Cobb experienced neo-enslavement as a sharecropper just after the period of enslavement: whites still used scripture shifting from slave codes to black codes to abuse blacks and control black labor.

Conversely, the beauty of scripture is it is elastic, exhorting and lifting people up: the Word is applicable in 702 BC and 2025. Consider a 20th and 21st century care of creation in the scripture looking to Leviticus 25:1 and 5:

> The LORD spoke to Moses on Mount Sinai, saying … You shall not reap the aftergrowth of your harvest or gather the grapes of your unpruned vine: it shall be a year of complete rest for the land.

Explicitly, the passage instructs readers and practitioners of any generation to till fields for gain while also respecting and caring for the land by leaving it fallow for a year. A farmer prudently cultivates and harvests the land, allows it to be fallow for a year for greater productivity in the future, embracing creation care. Interpreting more broadly, God requires we be caretakers of the planet.

The text is silent on specifics on who is to leave the land fallow. Yet it can be inferred that poor and working-class people did most of the hands-on labor of tilling the fields of cotton and vegetables for profit and personal consumption for the masses and the elite. Today, one can conclude this is an admonition for everyone whether a farmer in the ancient world or a family recycling plastic. In addition, someone had to actively decide to leave fields fallow. This is a call to be proactive, be an activist leaving the land to rest, which extends to alleviating climate change. Plant indigenous plants that attract pollinating bees in yards using the small plots judiciously. Rest the soil. Fertilize the soil. Broaden scriptural interpretation, guaranteeing that marginalized people, including blacks in places like Jackson, Mississippi, have potable water. Again, scripture and its interpretation are elastic in the past and present for the greater good. Leaving the land fallow is a metaphor for responding to climate change by and for BIPOC.

Again, drawing from Leviticus 25:1 and 5, black creation of care puts people first and nature second; people first rejects dominion of nature – subdue and exploit the earth – expanding the interpretation to black bodies and care of creation. Ta'Nehisi Coates describes how whites abuse black bodies in his book *Between the World and Me* (2015). Expanding on this idea of black bodies, many whites continue to operate out of biblical dominion, controlling rather than caring for nature. Whites have exploited and destroyed black bodies including during the Middle Passage and enslavement as part of the control over nature. Whites have abused black bodies in the cotton fields continuing with 20th century with blacks' blood flowing into the streets because of police violence in most often in built urban environments. Many contemporary white environmental organizations are generally more concerned about the planet and non-human creatures, and less so marginalized people. These organizations often ignore the plight of endangered brown and black bodies, brown and black people. Whites treat post-colonized landscapes, and black and brown bodies with disregard. Such racism, specifically environmental racism, has echoed in the aftermath of hurricanes in Puerto Rico, the Bahamas, and New Orleans: people of African descent experience the environmental brunt of the aftermath of the devastation. With federal environmental justice initiatives rolled back, the work of all environmental nonprofits moving us forward is even more critical.[10]

Those on the edges of society – including the unhoused, immigrants, seniors, citizens of developing nations, and BIPOC – have less control compared to people in power with more control of the environment.

Conversely, those with various privileges are safely housed, comfortable in developed nations, and financially stable in the upper-middle and upper classes. They are exposed but better shielded because of layers of socio-economic and geo-political privilege. Ultimately, black creation care advocates for the vulnerable people trying to survive the obstacles of life including climate change.[11]

Ned Cobb, among other blacks, was emblematic of proto-black care. Cobb lived between 1885 and 1973, cultivating the land during the 20th century. He grew cotton on flat arable fields filled with neat rows in Alabama. I imagine his hands darkened by the sun rubbing gritty soil between his fingers gauging the tilth or quality of the soil. Belief in God surely gave his labor on the soil scriptural context. The title of Cobb's autobiography is *All God's Dangers: The Life of Nate Shaw* (1975) – he used the pseudonym Nate Shaw to protect himself against racist backlash when the book was published – indicating he was a man of God. His real name was Ned Cobb. Though Cobb was illiterate, he may have heard Leviticus 25:1-5, among many other biblical references to the environment read and preached by black pastors from the pulpit.

Cobb was a sharecropper in Tallapoosa, Alabama, where whites utilized and profited from black bodies in cotton fields. Yes, Cobb experienced all this and was historically a black Christian, a proponent of care of creation praxis, and his book central to theological thought about the environment. He modeled and told others to care for the land. Influenced by Christianity, Cobb tried to practice creation care though agriculture. He did not use creation care or stewardship language – being responsible for what God gives us – but that is what he tried to do in Alabama. His faith in God informed his care for people living off the land first followed by the environment, typical of black eco-theology. Cobb did so honoring God because he was bound by his faith.

Cobb, one African American, had to weigh both utility and care of creation. Again, consider creation care for ancient agrarian people. "The Sabbatical Year," highlighted in Leviticus 25:1-5, honors the earth, taking only what is needed and returning what was not to preserve the earth for further cultivation, reflections on creation care. This is also reflects the realities of humans living in a modern world on our endangered planet.

Cobb lived in the tension between conservation – utility of and profit from agricultural resources and preservation – care for the environment like safeguarding forests, oceans, and people. He went against his beliefs because of agricultural practices and economic necessity. In Cobb's farming world, he knew that spreading arsenic poison, a form of conservation, killed boll weevils to save the cotton, a cash crop. At the same time, arsenic destroyed parts of creation including people's health like Cobb and his family downwind of the poison. Desiring untainted and healthy air, soil and water, Cobb experimented with organic or preservation methods turning from poisoning. He urged his children to pick weevils by hand from the cotton bolls.

Ultimately, Cobb pivoted back to conservation using arsenic for profit and a living, feed his family, and survive. Conservation is one aspect of dominion: exploiting resources or creation, whether natural or cultivated, as one sees fit to often to the detriment to plants to animals to people. Ironically, those who are marginalized, like Cobb, often have only one choice: abusing the environment to survive mirroring whites. Marginalized people mean marginalization of the earth.

Isabella Bomfree, born in 1797, was another black voice and perspective on creation. From birth, she was enslaved in the Dutch town of Swartekill – today the location is Rifton – in rural southern New York. Later in 1843 as an adult, Bomfree, experienced a religious transformation, having previously been on a spiritual journey attending church revivals. Isabelle Bomfree changed her name to Sojourner Truth, a reflection of the transformation. In the name of God, she traveled, speaking truth to power, justice to injustice. As an abolitionist and public speaker, Truth is best known for "Ain't I a Woman," her intersectional speech against gender and racial oppression.

The Narrative of Sojourner Truth (1828), her story from enslavement to freedom, is primarily about her broader Christian experiences, while dotted with theological reflections on creation. Truth, much like St. Francis of Hippo, appreciated nature through her faith in God. As important, she, like Ned Cobb, came from a different time than Francis. The difference was Truth, experienced racism and sexism.

Out of Truth's experiences, she sought solace in God's creation. Her musings are critical because the dominant voices about creation and by extension contemporary care of creation continue to be white. Olive Gilbert, who took notes from Truth and transcribed the narrative, described Truth's thoughts as theological. By extension, such insight makes sense concerning Truth's reflections on creation. She confirms this looking back to "our first parents, Adam and Eve, in the beginning, when they came from the hand of their Creator."[12]

Truth's appreciation of creation or nature was an escape from whites and enslavement, and a path to God. She learned from her enslaved mother the latter who after working, "would sit down under the sparkling vault of heaven" with her children noting that "the only Being that could effectually aid or protect them." Her mother assured her children that God "hears and sees you." Under the stars, part of God's creation, they worshiped God. Truth's mother added: "'He lives in the sky,' … 'and when you are beaten, or cruelly treated, or fall into any trouble, you must ask help of him, and he will always hear and help you.'" She led Truth and her sisters in the Lord's Prayer, and

> would point them to the stars, and say, in her peculiar language, 'Those are the same stars, and that is the same moon, that look down upon your brothers and sisters, and which they see as they look up to them, though they are ever so far away from us, and each other.'

Truth's mother introduced her children to a cosmic creation of God's making.[13]

Her mother's words and actions left a generational legacy to the children. Truth's mother fashioned her own worship in creation "under the open canopy of heaven, speaking very loud, she should the more readily be heard; consequently, she sought a fitting spot for this, her rural sanctuary." Truth offered "daily orisons" or prayers on:

> a small island in a small stream, covered with large willow shrubbery, beneath which the sheep had made their pleasant winding paths; and sheltering themselves from the scorching rays of a noon-tied sun, luxuriated in the cool shadows of the graceful willows, as they listened to the tiny falls of the silver waters. It was a lonely spot, and chosen by her for its beauty, its retirement, and because she thought that there, in the noise of those waters, she could speak louder to God without being overheard by any who might pass that way. When she had made choice of her sanctum, at a point of the island where the stream met, after. Having been separated, she improved it by pulling away the branches of the shrubs from the centre, and weaving them together for a wall on the outside, forming a circular arched alcove, made entirely of the graceful willow. To this place she resorted daily, and in pressing times much more frequently.

Truth said she had

> talks with God were perfectly original and unique, and would be well worth presenting, were it possible to give the tones and manner with the words; but no adequate idea of them can be written while the tones and manner remain inexpressible.[14]

Truth's cultural experiences framed creation care as was true of the novelist Toni Morrison. Morrison lived from 1931 to 2019, and as an adult she authored world renowned novels and was awarded a Pulitzer Prize and the Nobel Prize in literature. Morrison elevated black agency in fictional worlds with her protagonists facing racism and sexism throughout her novels. In her novel *Beloved* (1987), Morrison layered a creation story for black people with the white world on the edges as spectators or supporting characters. Still whites remain and at times execute racist incursions into ephemeral places of safety knitted together by Morrison.

One central character is Beloved. Central to the novel is Sethe who commits filicide against her daughter Beloved. In Sethe's mind a mercy killing spared the girl-child the cruelty of enslavement. Going deeper into the novel, perhaps Beloved's and Sethe's experiences were the first creation of Adam and Eve, and the second creation after the flood of Noah, respectively. Water is central in African spirituality through creation stories and redemptive in baptism introduced through the Christianity of whites to blacks. Consider whether Morrison drew on both African and white influences in re-creation for both Beloved and Sethe, daughter and mother. The restorative power of water moves both women from the violence inflicted on them by a white slaveholder to healing by the end of the novel.

Morrison was probably familiar with God creating water on the second of seven days in the biblical creation narrative. God began to outline creation in the first pages of the Book of Genesis. In Chapter 1 verses 1-2, God created "the heavens and earth" and "a wind from God swept over the face of the waters." Genesis 6-7 continues the theme of water:

> And God said, 'Let there be a dome in the midst of the waters, and let it separate the waters from the waters.' God made the dome and separated the waters that were under the dome from the waters that were above the dome. And it was so.[15]

Through water, Morrison reclaims creation for blacks primarily through Beloved, Sethe's dead baby daughter, came back to life or reborn. In three instances and within this context, Morrison weaves water into a series of scenes in the novel.

A nod to the first creation, Beloved was born or reborn coming out of a stream to the bank in new well-tailored clothing. She was grown but her lungs hurt like they were new as if testing them out for the first time. Almost as if soaking from a baptism, a renewal, Beloved was "sopping wet and breathing shallow." Like a newborn, "she spent those hours trying to negotiate the weight of her eyelids." Mirroring the second day of the biblical creation, Beloved sat with the breeze blowing, drying, and wrinkling her dress. Morrison says, "It took her the whole of the next morning to lift herself from the ground and make her way through the woods past a giant temple of boxwood to the field and the yard of the slate-gray house."[16]

In the second creation, Morrison invoked Noahic image of water when an adult Beloved reappeared, the ghost of the murdered baby made corporeal. Subconsciously, Sethe recognized the second Beloved, the former reacting physically. Sethe urinated, breaking her water as if once again in labor. Through a second birth or creation, both symbolic of the birth first child out of amniotic fluid and manifested physically a second time, as an adult Beloved. Morrison says of Sethe,

> She never made it to the outhouse. Right in front of its door she had to lift her skirts, and the water she voided was endless. Like a horse, she thought, but as it went on and on she thought, no, more like flooding the boat when Denver was born. So much water Amy said, 'Hold on Lu. You going to sink us you keep that up.' But there was no stopping water breaking from a breaking womb and there was no stopping now.

Sethe remembered giving birth to her daughter with the help of Amy, a stranger. As was true for Noah, the waters seemingly and endlessly flooded the world. The waters of birth was gone reflecting water receding in God's second creation, a renewal.[17]

Are we in a loop of destruction that after much pain will reap the renewal? Will we escape the flood to dry land like Noah? Will Sethe voiding her water, represent better times after enslavement from enslavement and the life of the planet?

Today, in the context of Noah, Beloved, and Sethe, humanity worsens the climate resulting in destructive waters, rising temperatures, and cataclysmic fire destroying more landscapes and human lives. From the perspective of black eco-theology and black liberation theology, BIPOC, including people of African descent, experience the implications of climate change more deeply because of lower socio-economic status, the metaphors of water: the Noahic floods and Sethe urinating.

Octavia Butler, renowned science fiction author and black futurist, was also prescient about climate change for those on the lowest rungs of society. In her novel *Parable of the Talents* (1998), she names climate change in her fictional rendering, saying,

> The climate is still changing, warming. It's supposed to settle at a new stable state someday. Until then, we'll go on getting a lot of violent erratic weather around the world. Sea level is still rising and chewing away at low-lying coastal areas like the sand dunes that used to protect Humboldt Bay and Arcata Bay just north of us. Half the crops in the Midwest and South are still withering from the heat, drowning floods, or being torn to pieces by winds, so food prices are still high.

The hope is for marginalized people and creation, looking to spirituality and religion. Butler says:

> The warming has made tropical diseases like malaria and dengue normal parts of life in the warm wet Gulf Coast and southern Atlantic states. But people are beginning to adapt … There are fewer diseases that result from bad sanitation, spoiled food, or malnutrition. People boil the water they drink in cities where there's a problem and in squatter settlements with their open sewers–ditches. There are more gardens and old-fashioned skills in food preservation are being revived.[18]

Butler's novel highlights implications of climate change in the context of founding Earthseed, a religion drawn from Christianity, and more specifically the biblical "Parable of the Sower." Much like different people making different decisions about sowing seeds on hard or good ground, people and nations make decisions about creation care in response to climate change. Nations and individuals continue to choose infertile hard ground, and unless a black eco-theology of care of creation is articulated and more importantly practiced, disproportionately BIPOC will continuen suffer and die. Practical theology of black care of creation can help make the shift to save people first and then the planet.

Notes

1 Leviticus 25:1-5, NRSVUE; Theodore Rosengarten, *All God's Dangers: The Life of Nate Shaw* (New York: Vintage Books, 1989), 221–223; and In the tradition of Octavia Butler, *Kindred* (New York: Doubleday, 1979).

2 Steven Bouma-Prediger, *For the Beauty of the Earth: A Christian Vision for Creation Care* (Grand Rapids: Baker Academic, 2010), xii.

3 Genesis 2:15, NRSVUE describes keeping creation and the NIV "taking care of it" and "Hebrew/Greek," "Biblegateway," https://www.biblegateway.com/passage/?search=Genesis+2%3A15&version=NIV&tab=interlinear.

4 Mark Rochelle Ferraren Renacia, "Contemporary Ecology and Augustine on Creation," *Estudios Eclesiásticos*, vol. 94, núm. 369 (junio 2019), 368, 369; 371 and St. Augustine, *Vol. 1: The Literal Meaning of Genesis* (Redbank, NJ: Newman Press, 1982), 42; and Saint Augustine, *Confessions*, trans. R. S. Pine-Coffin (New York: Penguin, 1961), Book XII, Section 7, 284–85.

5 Francis of Assisi, "Canticle of Brother Sun and Sister Moon," Ignatian Solidarity Network, https://ignatiansolidarity.net/blog/2015/06/04/canticle-of-brother-sun-and-sister-moon-of-st-francis-of-assisi/.

Chiara Mercuri, *Francis of Assisi: A Hidden Story* (Brewster, MA: San Damiano Books, 2019), 190–196 and Jon M. Sweeney, *St. Francis of Assisi: His Life, Teachings, and Practice* (New York: St. Martins Essential, 2019), 97–98.

6 Pope Francis, 90.

7 Pope Francis, *Encyclical on Climate Change & Equity: On Care for Our Common Home* (Brooklyn: Melville House, 2015), 91, and 133.

8 Ephesians 6:5, NRSVUE; Lynn White, Jr., "The Historical Roots of Our Ecological Crisis," *Science*, vol. 155, no. 3767 (March 10, 1967), 1287; and Roderick Nash, *The Rights of Nature: A History of Environmental Ethics* (Madison: University of Wisconsin Press, 1989).

9 Esther Bamitale Olumide, "A Comparative Analysis of Genesis Creation Story and Yorùbá Myth of Creation," (Kwara State University (Nigeria) ProQuest Dissertations Publishing, 2019), 1.

10 Ta-Nehisi Coates, *Between the World and Me* (New York: Spiegel & Grau, 2015), 35–36.

11 Melanie L. Harris, *Ecowomanism: African American Women and Earth Honoring Faiths* (Maryknoll, NY: Orbis Books, 2017), 20 and Delores S. Williams, *Sisters in the Wilderness: The Challenges of Womanist God Talk* (Maryknoll, NY: Orbis Books, 1993).
12 Sojourner Truth, *Narrative of Sojourner Truth* (Mineola, NY: Dover Publications Inc, 1997), 38.
13 Truth, 4.
14 Truth, 32.
15 Genesis 1:1-2 and 6-7, NRSV.
16 Toni Morrison, *Beloved* (New York: Vintage Books, 1987, 2004), 61.
17 Morrison, 62.
18 Octavia Butler, *Parable of the Sower* (New York: Grand Central Publishing, 1998, 2007), 79.

3 Sacred Burials

> All go to one place, all are from the dust, and all turn to dust again.
>
> –Ecclesiastes 3:20, NRSVUE

"Strange Fruit" was written by Abel Meeropol and recorded by Billie Holiday in 1935 is for 2025.

X hangs from a tree and he thinks of Billie Holiday's "Strange Fruit" in his last seconds. He looks out at the crowd of white people. Some are standing in excitement watching his life ebb away. Others are seated eating food they pull out of wicker baskets; and here he hangs swinging from a tree, strange fruit. As life ebbs, X thinks, "What a bad death. I will rot like fruit that is never picked until I fall from this tree to the ground bruised and forgotten." X's body is of no value. "Strange Fruit" is environmental context somewhere between lynching by whites and modern-day execution of blacks.

Today, blood on the urban streets. Blood on the ground. Blood on asphalt. Blood on concrete. The flowing blood remains a sacrilege, abuse, and destruction of black bodies by the hands of whites. Neglected. More bad deaths against the backdrop of asphalt, concrete, some trees, and small patches of grass and weeds.

In sacred memory of the dead, black bodies, buried and unburied, honored and dishonored: Africans captured who later died in holds and jumped from ships into the Atlantic during the Middle Passage; the enslaved, killed just because … Blacks lynched through the centuries … Blacks trapped by Katrina in the New Orleans SuperDome … The South Carolina 9 … Amadou Diallo … George Floyd.

The connection to sacred burial rites on sacred ground has been disrupted by violence because of racism. Though mostly black men and boys, black women and girls are part of a disconnect of burial rites short-circuited by bullets and killer chokeholds. These disrupted rituals are at the intersection of spirituality or religion, nature, racism, and culture.

DOI: 10.4324/9780429200700-3

Blacks are taking back what has been taken by whites. Nature and environmental injustice extend to black bodies and interments. Blacks caring for their dead, the bodies, even with violence, was resistance against oppression by whites often in natural places. Historically, blacks have lived with the racial inequity of segregated cemeteries, which at the same time reinforced a sense of black community; and for blacks including Africans, there are parallels with how whites mourn the dead. Some elements of nature in burial rites were in caves, the earth, images of animals, and water. Death is part of nature; nature is part of death. Death is not always treated equally among people.

On August 9, 2014, Michael Brown died in the streets of Ferguson, Missouri. He did not die from natural causes like old age. Instead, Darren Wilson, a white police officer, shot and killed Michael, an 18-year-old African American. Michael's death reflects countless moments of violence by police officers against black men and black boys in the United States. It was a bad death, as some Africans say.

Racist acts have long ended black lives too early and violently, while circumventing sacred and traditional funeral rites often in nature among and by blacks. Michael's black body was not treated with reverence; typically, humans – we are sentient – adhere to societal mores treating the dead with reverence through sacred acts. These mores have been shattered by racism, including a police officer shooting bullets in Michael's body. Michael's blood seeped from his defiled body to the unconsecrated ground below him. His body was desecrated.[1]

More law enforcement including SWAT converged on the street turning Ferguson into a police camp blocking people from getting close to Michael's body, in a built urban environment dotted with trees and patches of grass. The police committed a sacrilege against humanity, against Michael; against family, friends, neighbors, television viewers; and against nature.

Michael lay out in the open, his body uncovered in the summer heat for about 45 minutes from 2:11p and at 3:30p. In the street, the medical examiner then cleared the body to head to the morgue. Three autopsies followed, further ruining Michael's body, that of a young black teen.

The black community consecrated Michael's death when the white establishment did not. On Monday, August 24, 2014, Brown family, friends, and other grievers took Michael back from whites and racialized violence on that terrible day. Blacks did so in more formal, sacred, and liturgical ways through a Christian celebration of life and burial. It was mostly blacks who flowed through the doors of Friendly Temple Missionary Baptist Church, a black sanctuary in St Louis, Missouri. About 2,500 remembered Michael. Later at St. Peter's Cemetery in St. Louis, horses pulled his body encased in glass and all in white – drew Michael to his final resting place, a grave in the dark overturned soil of the cemetery.

As was true with Brown's mourners, funeral rites for blacks have long been simple to complex. For many it has been straight forward: a loved one dies of natural causes and at its simplest that person is buried followed by mourning. With internment, the earth, the simple turn of the soil cannot be separated from the complex holy acts of funeralizing and burial.

In more complex ways, Michael and other blacks – now ancestors, murdered violently, including at the hands of police officers – are viewed differently and negatively compared to a good death. Carol Beckwith says, "Africans say that when a spirit leaves the body, it is disoriented and clings to the places and people that it knew in life." Under negative circumstances, can ancestors intercede between God and recently deceased loved ones to ease the bad death into a settled afterlife? Does a bad death leave these ancestors with no place, bound and untethered just beyond earth, and the living who may forget loved ones or hold onto negative memories of these ancestors? Taken from black including African perspectives, a sad legacy of violence lingers, held by the living who mourn in a multitude of ways the racist violence against black men. These ancestors, whether through a good or bad death, ultimately still have a place with "the creator god."[2]

The respect due the dead, due to Michael, was dismantled and transformed by racist violent acts. There are countless variations of this sacrilege against black bodies, a disruption of funeral rites that progress to natural places like caves and the earth through the centuries, continuing even today. Still through debilitating racism, people of African descent have prevailed and excelled taking back brutalized black bodies through communal ritual acts, based in cultural norms connected to nature.

The responses of local people flooding the streets in Michael's memory were variations on historical and modern church testimonials in actions and words. They reacted to the sacrilege against so many blacks including Michael. The people demonstrated against the injustice of his violent death in the streets for about 11 days where Michael's body once lay in the heat. Mostly blacks organized peaceful protests and civil disobedience against the government including law enforcement. In doing so, they sanctified or consecrated his death, his ruined body, trying to balance justice with whites brutalizing blacks, erasing the latter's humanity. Blacks honored the dead under brutal circumstances, on the backdrop of an impoverished urban setting of asphalt and cement in Ferguson. All go to one place, to dust. All are not equal though in dying and death on this earth. Yet there is equality in finally being dead and going to a good afterlife or heaven.

The black church followed the demonstrators, finally able to formally remember Michael's death and consecrate his body with a church service and a burial. Though the sanctity of human life and sacredness of death was diminished by racism and by extension murder, blacks continue to invoke holiness in funeral rituals as was true of Michael's homegoing, part of many positive black church traditions. Blacks, throughout history and the world have honored their dead even when dehumanized by those in power. More importantly, people of the African Diaspora embraced their culture including rituals honoring the dead. These practices varied from geographic region to geographic region, and ethnic group to ethnic group. In doing so in many ways, they turned and returned to natural places like the earth to bury their dead. These were acts of liberation in the context of oppression, religion, and the environment.

Looking to the past, blacks, really all of us, began in prehistory with our origins in Africa, the cradle of civilization. Did early hominins honor their dead while reflecting on a deity, pondering an afterlife, and consciously doing so in natural settings? Early hominins, including Neanderthals probably had funeral rites. Conceivably these hominins were sentient, reflected in their connection to their dead ancestors; they did not leave their dead out in the open for any animal to scavenge. Instead, they intentionally left their dead inside caves, natural places, as far back as 160,000 years. David Lewis-Williams and David Pearce argue much the same about "megalithic tombs:" "They were places where the dead were revisited and where people retained long-term relations with them; they were religious and social foci."[3]

Michael Kerrigan adds:

> That the burial sites were monuments to thought, embodiments of the ideas by which these supposedly primitive people organized their world. In their orientation, in the causeways and kerbstones over which they were approached; in their external shape and their internal layout, the distribution of their different chambers; in the presence of niches, marks or more elaborate inscriptions—in all these aspects, the Neolithic peoples mapped out an aesthetic and a sense of order.

Carefully laying the bodies out on the surface or floors of caves or digging to bury the dead reflected nascent spiritual self-awareness.[4]

Hypothetical, theoretical, yes, yet some context looking ahead to homo sapiens or modern humans with higher thinking and learning compared to early hominins. As a result of such thinking, modern people increasingly turned to more complex funeral rites, part of which was in natural places. Homo sapiens including Jews and Christians from ancient times, funeralized and buried their dead so such rituals were and are not exclusive to blacks.

Christianity, the principal religion of blacks in the United States, is based in part on Judaism. Sacred texts, the Old Testament Bible, and the Torah, connect the cycles of death and life in nature. Ecclesiastes 3:20, (NRSVUE) says, "All go to one place, all are from the dust, and all turn to dust again." We die, become dust, returning to and enriching the earth: from life to death to dust, people practiced religious rites honoring the dead often placing them in natural places.[5]

Ancient Middle Eastern practices included burial rites in nature. Jews used openings often in natural caves on cliffs in hills. The alternative was digging graves in the sands of harsh shifting desert landscapes and being unable to find their dead later. These caves held beloved families who they laid to rest. For example, according to Genesis 23, Hebrew patriarch Abraham purchased the cave of Machpelah, when Sarah, his wife, passed away. The acquisition ensured that generations of Abraham's family would be entombed there in nature.

In another example from John 11:17 (NRSVUE), Lazarus, a man Jesus miraculously brought back to life, was probably buried in a sepulcher or a space carved out a natural stone face to lay out the dead. Jesus was buried in a similar way. In these instances, following the cultural norms of burial or interment, families cleaned and wrapped the bodies of loved ones in cloth, probably linen; the bodies were quickly entombed before sunset, and the entrance of the cave sealed with stone. Bodies were laid to rest so quickly to honor the dead and send them quickly into the afterlife. Humans decompose quicker in the heat so the services and internment happened quickly after death. Loved ones returned about a year later to gather the bones and placed them in an ossuary, a box like contemporary urns that hold cremains.[6]

Shifting the geography and timeline to Europe during the late Middle Ages, the second plague, the Black Death or the Bubonic Plague decimated perhaps a third of the population with an estimated 25 million deaths. As a result, Catholics, the primary Christian faith in Europe, were forced from ceremonial burials to mass graves for practical purposes during the mid-14th century. They did so because leaving bodies unattended without quick burials led to the spread of other pathogens like gastro-intestinal infections from feces and attracted rodents with fleas that spread the plague even further. Sadly, dead beloved mothers, fathers, children, and friends were unceremoniously wheeled off in wooden carts to mass graves. People dug large pits for countless dead, and loved ones mourned without the closure of a formal religious funeral with solemnity and mourning for individual burials. At least for a time, gone were single burial spots with chiseled headstones to help memorialize the dead loved ones in sacred soil.

Further along the historical timeline, Europeans including the English returned to more formalized burials. Protestantism was the denomination of the state and dominant religion in England from the late 16th century, with Henry VIII disavowing the Catholic Church. Starting in the early 17th century, English settlers – in what would become the 13 colonies – practiced Protestantism including Puritanism, as they rejected the state church in England and the state rejected people who practiced other sects of Protestantism.

Without the tragic circumstances of mass death during the plagues, and pivoting from some European traditions, the English who landed in the Americas, reframed funeral rites around community including the family, relying less on the European emphasis on Anglican and Catholic priests at the center of ceremonial grief. Kerrigan says,

> As soon as he heard the news of a person's passing, the town sexton tolled the bell: nine chimes for a man, six for a woman, and three for a child … Neighbors gathered at the house to comfort the bereaved and help laying out the corpse—washing it, trimming untidy hair and, if an adult male, giving a final shave. It was a close-knit neighborly affair, not yet professionalized by morticians of today.[7]

During the late 19th century, funeral services became professionalized through funeral homes and funeral directors. As part of this professionalization, funeral homes embalmed the bodies giving families more time to bury their dead. Consider an example, and advantage to embalming: an adult child traveling cross country continues to have more time to arrive for funeral rites and mourning with family.

Arguably, the very act of finding places to inter, bury, or cremate one's dead can be traced throughout the history of North America, specifically the United States. Soil, in graveyards continued to be very human entry points to religion, spirituality, and nature. These acts of burials bring humans through their dead closer to nature.

To paraphrase W. E. B. Dubois, blacks were caught between a double consciousness of being both African and American. They retained Africanisms from Africa but also assimilated into white culture to survive. The duality was through their faith, spirituality, and religion.

Africans long practiced funeral rites forged primarily out of African cultures until European invaders, imperialists, and colonizers arrived. Shifting to contemporary Africa retaining some historical influences, the variety of rituals around death including burials are as numerous as there are countries, ethnic groups, communities, and languages on the continent. Still there are some generalizations. Worldwide and throughout history "funerals are considered the last transitional rite" and for Africans "a means of formally introducing an individual into the world of the spirits with "the worlds of the living and dead are equally real." Generally, the body is laid to rest often in the earth, and the spirit moves into the afterlife.

Africans turned to nature prior to internment. Some examples include animal sacrifice in Mali and Uganda. In Mali, prior to a funeral, "Dogón high priests sacrifice certain culturally significant birds and animals each day for six weeks." In addition, "the pastoral Bangalore of Uganda slaughter all the full-grown bulls ... sprinkling their blood around the house and over the bereaved family as a gesture of purification," honoring a deceased man who was affluent and influential.[8]

Most Africans, including the Ashanti, quickly buried their dead within a day or two because of the extreme heat knowing that the bodies quickly decomposed. Elements of nature, specifically water, have been used for caring for the dead in the afterlife. For example, the Ashanti attend to a loved one with deference to avoid thirst in the afterlife. For the burial, "coins are tied to a corner of the burial cloth so that they will be able to buy drinking water on their arrival on the other side." The mourners also bathed the body three times with relatives on the maternal side cleaning the right side and those on the paternal side cleaning the left side. In doing so, mourners believed they were blessed, eliminating illness, improving health, and increasing fertility. They used rum, part water, to clean their hands to remove residue of the body, "pouring a little down the throat of the corpse to 'keep it happy' and delay the process of decomposition."[9]

In Africa, options for burials have often been set in nature. The Surma of Ethiopia, quickly completed burials in a day again because of the heat. They coated the body with milk as a blessing. The Surma placed the body upright in the middle of a household's compound. Hottentots in South Africa "bury a body flat on its back in a deep hole, with its head turned to the west, marking the head of the grave only with a small stone." In doing so, the deceased would face the setting of the sun at the end of the day, perhaps symbolic of the end of life. The Bantu of East Africa, "seeking to please the deceased, dig a room underground chamber, when the body can move easily should it wish so." In Bantama, one of the burial places of the Ashanti kings was near a sacred tree – where loved ones who became ancestors lived – was predictably near the mausoleum and the shrines.[10]

After burials, some groups continued to honor the dead. Ethnic groups in Ghana, Burkina Faso, and Sudan, "mixed and placed millet, water, and chicken blood at the corners of fallow fields, as a consecration of the soil for planting and honoring the deceased."[11]

A bad death in Africa continued to be a bad death into the Middle Passage and the Americas. From the 1500s to the 1800s, European seamen kidnapped and transported about 12 million Africans, most from West Africa. The trip was a horrifying odyssey at sea, trapping and chaining Africans below deck in the holds of ships. Travel took about a month and a half. European seamen ultimately sold Africans in the Americas.

Europeans sailed slave ships that were built environments transformed and that nature took back. The Europeans constructed holds, dark cramped unventilated spaces, that were large Petri dishes, a putrid organic stew that was toxic and deadly to many Africans: vermin like rats and mice scurried over bodies carrying disease; with little or no food and water many starved; and diseases like smallpox, dysentery, and scurvy sickened and often killed Africans. Laying side by side, they excreted vomit, sweat, feces and urine into the hold. It is estimated as many as two million died in the Middle Passage. These devastating physical conditions led to many bad deaths.

Africans died in another way as they naturally became depressed in some cases committing suicide. Some jumped from the ship into the waters to die rather than endure the terrible conditions on the ship. Those who jumped from the deck into the ocean were often women with their children and babies; Europeans allowed those considered docile on deck for the sunlight and fresh air. African men who somehow escaped the hold jumped too. Others gave up hope, gave up eating and drinking, and gave up life. In doing so, Africans retained their dignity making choices and took back their freedom. They rebelled.[12]

There were spiritual dimensions to Africans taking their lives by suicide. Choosing graves in the ocean and hold by suicide was a form of rebellion against white oppression, a hope that they would be returned to the land of Africa, a heaven of sorts. Suicide and rebellion were chances at a good afterlife, better than life in the lice and rat-infested holds:

> On the slave ships Africans frequently turned to death, either by forced starvation, by starving themselves, or by leaping into the ocean. Some believed that if they were killed they would return again to their own country. In other words, many of the slaves may have taken solace in the thought that after death, they would return to Africa.[13]

Others died at sea when a ship sank in a storm or was attacked by another European country. In 1738, the Dutch ship Leusden sank off the shores of Suriname in South America. The white crew escaped leaving about 700 slaves in the hold with no way to escape. The Africans died terribly drowning. They held their breaths trying to keep water out and air in ... they swallowed and breathed in water ... they could no longer breathe and died.

Again, according to African traditions, some were generally bad deaths – suicide, murder, drowning, and death by disease – because of whites holding blacks bondage. These gruesome ends were unlike say a natural death from old age surrounded by loving family and friends. The gift of traditional burial rituals was torn from family and friends who were left behind in Africa. Perhaps worse, those shackled in the hold of ships could not ritualize the death of the man or woman – many strangers to one another but still African – lying dead next to them. African culture, sacred rites were stolen along with the humanity of those from the shores of the African continent.

Those Africans who survived were limited to practicing or could no longer practice traditional dying and death rituals. These practices included sacrificing an animal for the blood as an offering to deities on behalf of the dead or washing the deceased with milk as a process of cleansing the body. Perhaps for the sake of trying to continue rituals Africans used other sources of blood for sacrifice to reinvent funeral rituals in the hold of those ships. They could still speak so might have used the oral portions of funeral rites. We may never know as Africans relied heavily on oral traditions rather than the written word, and whites barred most of the enslaved from reading and writing.

For Africans who were left behind, they lost the chance to ritually bury their dead. That included lovingly bathing the bodies of family with water. With rites, including those tied to nature, those in Africa could never fully mourn people who were stolen from them.

"The watery grave of the 'middle passage' ... to land bound slave cemeteries" dramatically transformed the lives of people of African descent in white dominated America. Africans who survived the Middle Passage, retained Africanisms while acculturating to white American society. As a result, people of African descent were shaped by social, cultural, political and economic hybridization because of emotional, verbal and physical abuse and violence by whites. Whites nefariously invoked the Bible in many ways including "slaves obey your master" as part of the emotional and verbal abuse. Through this grinding process, whites made too many choices for blacks.[14]

Blacks also made their own choices snatching back some autonomy in their new communities in the Americas. Again, and in many ways, Africans assimilated to adapt to the harshness of being enslaved and the new landscapes of the Americas. Blacks retained glimmers of African spirituality including reinforcing the importance of healthy black communities faced with racist adversity. They were resilient, surviving and excelling. Even with the horrific barriers of enslavement and whites brutally stripping away African identities and cultures, some Africanisms helped retain connection to nature through religion and spirituality.

As a result, blacks had to re-envision and retain remnants of centuries of African rituals of dying, death, and burial in a world that segregated blacks. For instance, people of African descent continued to remember and revere the ancestors within the context of nature much like Africans. Some layered Christian beliefs – saints as ancestors and Jesus as the ultimate ancestor with African spirituality. Through dual lenses of blackness and whiteness, the slaved survived, endured, and persisted burying their dead though the institution of slavery did not make it easy. Blacks rebelled to creatively honor their dead.

Light ... Darkness ... Sunup and sundown ... Burial into the earth – all were aspects of nature in the context of dual African and European identities. During enslavement and as an example of the duality, whites forced blacks to shift from daylight to nighttime funeral services in the context of nature. Sadly, the enslaved worked from sunup to sundown, daylight hours with little room for personal time but Sundays, another impact by the dominant culture. As a result, many of the enslaved buried beloved relatives and friends, their ancestors in an African context on American landscapes, after an arduous day of work in the slaveholder's fields and homes. Kerrigan says,

> It shocked white Americans—and frankly frightened them—that slaves held their funerals at night, singing loudly as they bore the deceased to the grave by the flickering light of torches. While this practice no doubt reflected different customs, traditions and attitudes that they had brought with them from Africa. It stemmed too from the fact that the hours of darkness were the only ones in which they had any degree of autonomy.

African entwined the cultural influences song and rhythmic movements with Christian burial. Blacks did not let the darkness of night stop them from properly burying their dead in racialized environments and the confines of laboring for free for whites.[15]

The spatial racial socioeconomics of burial sites for the enslaved was also decided by slaveholders. For example, from the mid-18th to the early 19th centuries, President Thomas Jefferson built his sprawling Monticello plantation including racialized burial places for whites and slaves outside of Charlottesville, Virginia. Jefferson, like most slaveholders, forced the enslaved to bury their dead, much like rudimentary slave cabins, both out of the sight line of whites. In addition, whites limited black attendees to two people at the funeral, where black cultural practices called for numerous people as communities to ceremonially bury the dead. Whites feared black numbers would become an insurrection. Blacks were even forbidden crudely made grave markers, a way to ritualize the location and continue to remember the ancestors. Greg Melville says,

> Burial grounds themselves were usually set far apart from the plantation house, on a patch that couldn't be farmed, grazed, otherwise used for profit – out of sight of visitors. The locations of these spots were almost never put into any written record. This is why we only know the locations of a tiny fraction.[16]

African spirituality, Christianity, and geography intersected with burials in the anti-slavery narrative of Harriet Jacobs, a runaway slave. She was an African American woman who had memories of a cemetery where funerals might have also taken place at night. Her pseudonym to protect her identity as a runaway slave was Linda Brent and she was an abolitionist. As part of her role resisting and exposing the horrors of enslavement, she wrote and published *Incidents in the Life of a Slave Girl* (1861).

In one passage of the book, Jacobs bridged an African and American experience, reflecting on her ancestor with a Christian cemetery, which was surely a segregated space for blacks burials. Jacobs described a spiritual moment that was solemn and sacred. In the context of African ancestor veneration, she reminisced about standing by her parents' grave near a vacant church falling apart set in nature in North Carolina. She further details,

> A black stump, at the head of my mother's grave was what remained of a tree my father had planted. His grave was marked by a small wooden board bearing his name, the letters of which were nearly obliterated. I knelt down and kissed them, and poured forth a prayer to God for guidance and support in the perilous step I was about to take.[17]

Jacobs also invoked the ancestor Nat Turner, one of the most famous insurrectionists who led a rebellion to free others who were enslaved in Southampton, Virginia. Finally, she made another connection to the ancestors saying her own father spoke to her telling her to run from enslavement in the South and to freedom in the North.[18]

All in all, these segregated cemeteries were set in nature: some wooded and all providing sacred soil to bury the ancestors. Jacobs in her own way remembered the outdoors: the woods, the waning light. and a bird sounding where her ancestors were buried.

Others, like Jacobs, revered their ancestors in many ways through burials as place both literally and figuratively. Some of the enslaved literally or physically dug graves "along an east-west axis" to avoid violating "the West African strictures against burying a person at the figurative 'crossing in the world'" or the crossroads.[19]

African religious burial beliefs meant in part standing at the crossroads of being snatched from the land of Africa and enslaved in the Americas, and claiming a new place, new earth in a racially contested space. The belief in these crossroads carried over into the Americas as a spiritual safety net. In African culture, the crossroads were the east-west and south-north with both lines crossing at the center at both axes. It was the spiritual and metaphorical tissue that connected the worlds of the living and the dead. Africans did not want to figuratively be left at the nexus or the crossroads between the living and the dead, life and the afterlife. No one wanted to be stuck in-between unable to fully be alive with family and friends or dead to be among ancestors. The light of the day-time sky might have figuratively represented Eshu, the African god of the cross roads. The living supported the dead or ancestors moving from the earthly world and ancestral afterlife. The crossroads meant life and death, beginnings and ends, birth and burials.

As for the east-west burial, the feet of black Christians one would be at the west, giving the dead a sightline of the rising sun coming up in the east.

Perhaps, the rising sun meant hope, guidance, and revelation. Another perspective of looking to the east, the "rising of Judgement Day; otherwise, the person is in the crossways of the world." Sinners buried with their feet to the east moved their sight line from seeing the sun, caused them harm and was opposite in direction of burials of good blacks with their feet to the east.[20]

Most of the enslaved Christians had headstones that faced east as was true of the heads of the dead facing the sun, the light. John 8:12 (NRSVUE) uses the metaphor of Christ being the "light of the world," a connection to the sun. Even in end times, there is light perhaps in connection to the East. Revelation 22:5 (NRSVUE) says, "And there will be no more night; they need no light of lamp or sun, for the Lord God will be their light, and they will reign forever and ever."[21]

After the enslaved were freed, white racism ensured that blacks and whites remained in segregated cemeteries: sacred spaces remained separate and unequal. For example, in 1852 Savannah, Georgia, whites opened a separate section for black free people called Laurel Grove Cemetery South. Whites buried their dead in the Laurel Grove North, which was well-appointed and beautifully landscaped, filled with well-kept lanes in the understory of oaks, along with ornate expensive statues and gravestones. Further on the timeline and in contrast to whites, the space, the place that was Laurel Grove South was south of an interstate and was far sparer and more understated with narrow lanes and much less landscaping including brown dry grass. Melville described the southern cemetery: "Its beauty instead radiates from its simplicity and closer connection to the earth." Though described as lovely, Laurel Grove South had less amenities and aesthetics because of segregation imposed by whites.[22]

In addition, black families were forced to bury their dead in lower ground vulnerable to storms, flooding, and damage, more inequity, while white graves in the north were on higher ground, a privilege. Yet, more recently there has been an environmental justice of sorts: though many black cemeteries have been or left in disrepair, the dead were ultimately cared for to a degree. Melville says,

> Crews removed the occupants from the old Black burial ground downtown and reinterred them there, planting the small, oddly shaped slate gravestones in orderly rows, but not necessarily atop the correct people. A small wooden sign that reads SLAVE BURIALS now marks the original burying spot. Within a decade, Laurel South grew to fifteen acres and expanded outward to more than double that size in the twentieth century.[23]

Today, the inequity between whites and blacks continues in many black burial grounds filled with overgrown weeds and in disrepair. In 1993, the Missouri River tore through the small town of Hardin, Missouri. The water destroyed homes, grain bins, barns, along with their utility grid including electricity and water. Much of the cemetery was on "a segregated acre where blacks were buried decades ago, many in unmarked graves. After the flood, the coroner and others marked newly dug graves for identified bodies but placed others in huge mass graves." In 2005, Hurricane Katrina was a harbinger of escalating climate change, sending caskets to float away and vaults forced open by the power of water. More recently, the caskets in a black cemetery in Brighton, Mississippi continue to be regularly submerged. Wright describes other challenges some of which are environmental: "Natural weathering, vandalism, neglect, abandonment, agricultural practices, encroachment by development, and industrial pollution all pose threats to these and other African American cemeteries." Black cemeteries that are bulldozed for development is a sacrilege against the remains and an insult to loved one's still living in earthly places amplified by racism. Racialized climate change and new construction worsened black cemeteries, upturned burial plots, and the emotional lives of the living watching the sacrilege of their dead.[24]

Today in the United States, most cultures and ethnic groups bury or cremate their dead to the detriment to people and the environment. The process of embalming continues to be a critical part of funeral rites that preserves the body for presentation at funerals and responding to human needs around time. The stench of decomposition is no longer a barrier. Unfortunately, the chemicals of embalming and metals in coffins leach into and poison the soil. In turn the chemicals and metals pollute the planet and negatively impact people.

Cremation continues to be another option used by funeral homes. Burning bodies as part of funeralizing goes back 17,000 to 20,000 years so is not new to modern times. Unfortunately, cremation requires the energy of fossil fuels for burning the bodies, pollutes the air with carbon monoxide and particulates, and helps to accelerate climate change. Improper disposal of ashes also pollutes the environment including the air and the soil.

Most people in the United States including blacks are turning to green or "natural cemeteries" and "are Regreening the deathscape." Green burials do not use chemicals for formaldehyde in embalming or cremation that pollute the air with sulfide oxide, and coffins made of metals including bronze, and vaults including concrete and plastic, which seep into the ground after burial. No vaults of cement. Biodegradable shrouds and urns. Such green, natural, or sustainable burials are increasingly becoming part of end-of-life conversations.[25]

In closing, Suzanne Kelly asks two critical questions with implications for communities of African descent:

> How are communities of color, in particular, and other disenfranchised groups not now well represented in movement building making sense of green burial in terms of their own histories? And in what ways will multiple cultural and community perspectives come to shape the movement of grassroots work continue to grow?

These questions are foundational to caring for the planet no matter the ethnicity.[26]

Notes

1 Suzanne Kelly, *Greening Death: Reclaiming Burial Practices and Restoring Our Tie to the Earth* (New York: Rowman & Littlefield Publishing, 2015), 24–25.
2 Carol Beckwith and Angela Fisher, *African Ceremonies* vol. 2 (New York: Harry N. Abrams, Inc., 1999), 237.
3 David J. Lewis-Williams and David Pierce, *Inside the Neolithic Mind: Consciousness, Cosmos, and the Realm of the Gods* (London: Thames and Hudson, 2005), 179 and Michael Kerrigan, *The History of Death: Burial Customs and Funeral Rites, From the Ancient World to the Modern Times* (Guilford, CT: The Lynn's Press, 2007), 41.
4 Lewis-Williams, 179 and Kerrigan, 41.
5 Ecclesiastes 3:20, NRSVUE.
6 Genesis 23, NRSVUE; John 11:17 NRSVUE; and Matthew 27:57-61, NRSVUE.
7 Kerrigan, 140 and 144.
8 Carol Beckwith and Angela Fisher, *African Ceremonies* vol. 2 (New York: Harry N. Abrams, Inc., 1999), 238 and 239.
9 Beckwith, 237.
10 Beckwith, 238 and Grace L. Chavis Butler, *Africa: Religions and Culture, With a Focus on the Ashanti People* (Pittsburgh: Dorrance Publishing Co., Inc. 1994), 214 and 215.
11 Dianne D. Glave, *Rooted in the Earth: Reclaiming the African American Environmental Heritage* (Chicago: Lawrence Hill Books, 2010), 44–45.
12 Roberta Hughes Wright and Wilbur B. Hughes III, *Lay Down Body: Living History in African American Cemeteries* (Detroit: Visible Ink Press, 1996), 3.
13 Hughes Wright, xxv.
14 Hughes, 3.
15 Kerrigan, 145.
16 Greg Melville, *Over My Dead Body: Unearthing the Hidden History of America's Cemeteries* (New York: Abrams Press, 2022), 50–51.

17 Glave, 46 and Harriet Jacobs, *Incidents in the Life of a Slave Girl* (New York: W. W. Norton, 2001), 80, 92–93.
18 Glave, 46 and Jacobs, 80, 92–93.
19 Wright, xxv.
20 Wright, 7, 19, and 20.
21 John 8:12, NRSVUE and Revelation 22:5, NRSVUE.
22 Greg Melville, *Over My Dead Body: Unearthing the Hidden History of America's Cemeteries* (New York: Abrams Press, 2022), 87.
23 Melville, 89.
24 Wright, 37 and 3–4.
25 Mark Harris, *Grave Matters: A Journey Through the Modern Funeral Industry to a Natural Way of Burial* (New York: Scribner, 2007), 2 and Kelly, 31.
26 Kelly, 62.

4 Resistance

> Slaves, obey your earthly masters with respect and fear, and with sincerity of heart, just as you would obey Christ.[1]
>
> –Ephesians 6:5, KJV

In the Exodus narrative, the pharaoh of Egypt ordered the midwives to kill all Hebrew newborn boys. He feared one of those boys would become a man, and one day usurp the pharaoh's throne. Shiphrah and Puah, Hebrew midwives, resisted and acted by defying pharaoh. The women chose to save the lives of the Hebrews, their own people, by rescuing the baby boys from death. That the king would punish the midwives was a given. Yet even with the certainty of death for their actions, the women rebelled by protecting their own in the name of God.[2]

Blacks have long looked to leaders like these midwives as models of resistance from The Bible. The Egyptians subjugated the Hebrews including the midwives Shiphrah and Puah. The women resisted in the tradition of the Hebrews who ultimately escaped from the Egyptians to freedom. Perhaps the women were among those who escaped. The Hebrews spent 40 years making their way to the Promised Land, known today as Israel.

Looking to the Promised Land, a metaphor for freedom, enslaved blacks invoked the Hebrews seeking freedom from the Egyptians. The Exodus narrative of enslavement, flight, and freedom – for both groups set against the backdrops of built and natural environments of pyramids and tobacco fields, respectively – is still potent today among blacks long marginalized through many centuries in the United States.

Even secular scholars default to powerful and engaging biblical narratives. Vincent Harding's *There is a River: The Black Struggle for Freedom in America* (1981) draws from the biblical Exodus narrative with chapter titles that include "Many Crossing Over," "Strange River in Canaan," "Beyond the North Star," and "On Jordan's Stormy Banks." Blacks and resistance theology are irrevocably bound to one another.

Nat Turner, a slave, resisted much like the Hebrews. Turner reflected a nascent black eco-theology expressing himself theologically through the

DOI: 10.4324/9780429200700-4

lens of nature in visions from God. His theology was based on Western Christianity that was imposed on him as a slave. Yet he made Christianity his own as a person of African descent perhaps also drawing on African cosmology. Through a wilderness experience theology of resisting oppression, Turner agitated. It is "the womanist theological supposition that black women's religious experiences are shaped by black women's interaction in black communities composed of both oppressed females and oppressed males."[3]

In 1831 Southampton, Virginia, Turner arguably led the most famous slave insurrection in North America. The insurrection was historically significant and continues to be part of the collective imagination of many in the US through the film "The Birth of a Nation" (2016).[4]

Turner believed that God gave him visions that were spiritually based and prophetic. The insurrectionist quoted scripture as part of his visions filled with images of nature. He did so in the tradition of Old Testament prophets. He believed he was called by God as Her/His prophet to lead a slave revolt. Given permission from God to rebel, Turner did so in the tradition of slave insurrectionists Gabriel Prosser and Denmark Vesey. Arguably these leaders all acted prophetically to resist enslavement incrementally transforming an oppressive institutionalized culture to one in which blacks were ultimately freed from enslavement.

As a black man enslaved by whites, Turner's spiritual journey of rebellion was filled with symbols of resistance he drew and interpreted from The Bible. Turner blended these symbols of nature based on daily life, the African ancestors, and Christianity. The influences from nature were probably based in part in African spirituality: life's routines, religion, and nature combined. Taken together, water may have meant many things to Turner: mundane yet necessary for drinking and bathing, purification in African traditions, and water baptism in Christianity.

Between African and American influences, Turner chose a better life, really a better afterlife; he probably knew death was inevitable when he organized and led a slave revolt based on the symbols of his faith from nature. He had two options. a tortured life as a slave or shorter life as an insurrectionist. Death could come early with the life expectancy of a slave shortened because of overwork, poor nutrition, disease, and physical abuse. Accepting enslavement meant being perpetually stripped of one's humanity and manhood. Alternately, Turner could resist the slave holders shortening his life further, while embracing a life of authenticity and meaning. The historical record tells us Turner chose manhood, resistance, and an early death. Ultimately, Turner believed God called him to rebellion "perfected by Divine inspiration," much like God directing the Hebrew midwives.[5]

On October 2, 1800, he was born into enslavement and Christianity in Southampton, Virginia. I imagine, as a child he felt the breeze from the ocean but rarely if ever played along the shore and in the water. Turner, like the others enslaved children, spent most of his time planting, tending, and priming tobacco out in the fields as he was stung by mosquitos. He was a black male child in a white world.

Yet Turner had another life of visions outside the punishing labor, denied the dignity of wages for farming cash crops like tobacco for export and subsistence produce like potatoes for the table. As a Christian, he had a rich interior life that was both intellectual and spiritual. Very early on, Turner learned to read from The Bible. He also had a great bond with his grandmother in their shared Christian faith. The visions and his life were that of a man of African descent forced to work as a slave.

The direction of Turner's life was influenced in many ways. He read The Bible certainly unusual since slave codes across the South forbade such activity. To quote the trite phrase: knowledge is power; and reading The Bible gave Turner powerful ideas to resist. Exposure to Christian tenets ultimately led to his revolutionary ideas, the foundation for Turner attempting to violently destroy the institution of slavery at least in his small world. Perhaps he envisioned young David taking down the Philistine Goliath with a sling and a stone – the slaves represented by David and the slaveholders by Goliath. In addition, while interpreting scripture on his own, Turner surely listened to and was influenced by the black and white pastors interpreting scripture in their preaching. Some surrounding Turner also acknowledged his spiritual development, as they watched him read The Bible and practice contemplation through fasting and prayer. As a child, some in the slave community predicted Turner would be a prophet destined for greatness. Other slaves affirmed his prophetic call as part of his manhood. Slaves who sat under his preaching and teaching at Baptist meetings referred to Turner as "The Prophet."[6]

These influences informed how Nat Turner might have interpreted The Bible very differently from whites. For Turner, that difference was based on being a person of African descent and enslaved. As a black Baptist during enslavement, Turner was probably immersed in the Old Testament including the Exodus narrative of resistance and freedom in the King James Version, typically read during the 19th century. In very different ways, whites adhered to and manipulated some scripture that suited them further securing their power as whites using Ephesians 6:5 (KJV): "Servants, be obedient to them that are your masters according to the flesh, with fear and trembling, in singleness of your heart, as unto Christ." Through Turner's actions, he rejected slaveholders' interpretations of this text; he rebelled against whites who enslaved him much like chattel: a horse, goat, or cow.[7]

Instead of adhering to whites' interpretation of enslavement, he might have reframed the exhortation to be good slaves. One law in Leviticus

18:3-4, (KJV) against sexual immorality could have more than one interpretation for enslaved blacks:

> After the doings of the land of Egypt, wherein ye dwelt, shall ye not do: and after the doings of the land of Canaan, whither I bring you, shall ye not do: neither shall ye walk in their ordinances. Ye shall do my judgments, and keep mine ordinances, to walk therein: I am the Lord your God.

Turner may have seen himself resisting slaveholders much like the Hebrews rejecting Egyptian law. He followed God's judgements, visions of nature from God countering the slaveholders holding his people of African descent in bondage.[8]

Turner was revolutionary – are not all prophets? – building the kingdom of heaven. With overlapping interpretations of the kingdom and heaven, Turner loosely quoted Matthew 6:33 from the King James version, typically read during the period: "I was struck with that particular passage which says: 'Seek ye the kingdom of Heaven and all things shall be added unto you.'"[9]

Such a kingdom could be interpreted in many ways. Prophets prophesied to make the kingdom on earth a better place by identifying and rectifying injustices. The kingdom of heaven also reflects spiritual experiences here on earth. For example, some welcome the indwelling of the Holy Spirit as a guide, experiencing heaven on earth; and ultimately after death, one truly becomes spiritual creatures in heaven. Turner believing all things were "added unto" him, which may have included this theology of heaven.[10]

Here on earth, Turner chose violence seeking the kingdom of God. He and other slaves physically attacked and killed whites attempting to be free, which might have been his heaven on earth. Even in death, surely Turner believed he was heaven-bound to the afterlife. Perhaps his mansion in the heavenly Promised Land awaited.[11]

In the tradition of Old Testament prophets like Isaiah, Ezekiel, Jeremiah, and Daniel, Turner resisted the status quo planning a slave insurrection. These prophets were spiritually connected to and were messengers of God. Turner understood the power of the prophet, the mounths, handsn and feet on behalf of God, in the black community. He embraced the role and encouraged others to follow him. Turner said, I "wrapped myself in mystery, devoting time to fasting and prayer," much like a prophet. He modeled himself on revolutionary prophets using environmental symbols to prophesy.[12]

Keeping Turner in mind, the prophet Samuel gives insight to the complexity of physical violence in scripture. As a prophet and leader of ancient Israel, the Philistines, one of Israel's enemies, threatened war. Yet in the spiritual realm, Samuel sought God's direction sacrificing an animal to seek answers from God. He prayed to God about the threat of the Philistines. Samuel was prepared to go to war. Ultimately, God appeared in the theophany as thunder frightening the Philistines away. Like Samuel facing the Philistines, white enemies surrounded Turner. He may have looked to prophets like Samuel for direction to fight back against whites; and the thunder came as Turner prepared and armed himself and others theologically for a physical insurrection.[13]

Keeping in mind prophetic traditions of The Bible, Turner understood four visions as God confirmed that Turner moved forward with the revolt. The first time Turner had a vision based in nature, he

> saw white spirits and black spirits engaged in battle, and the sun was darkened—the thunder rolled in the Heavens, and blood flowed in streams—and I heard a voice saying, 'Such is your luck, such you are called to see, and let it come rough or smooth, you must surely bare (sic) it.'

The battle between spirits represented the slaves and slaveholders.[14]

Who can say who was white or black as white has long been perceived as good and black as evil in Christian traditions? Perhaps as a black man, he interpreted the black spirits as good reversing the age-old stereotypes of good/white and evil/black, the latter often darkness in scripture. Looking to African Traditional Religion (ATR) including the perspectives of the Yoruba, evil is not broadly conceptualized as is true in Christian theology. Consider the Tree of Knowledge of Good and Evil in the Garden of Eden in Christian traditions that revealed good and evil to Adam and Eve. Their experiences were the theological foundation of good and evil. Alternately, in ATR, the focus is on groups and individuals being narrowly accountable for their evil actions compared the broader idea of evil in Christianity. He was surely placing the responsibility of the evils of enslavement on the slaveholders individually and collectively. Ultimately, Turner probably drew from both religious traditions; and as is true for many who practice Christianity, it is difficult to differentiate religious practices between ancient religious traditions from Protestantism.

Much like God defending Samuel and the Israelites with thunder against the Philistines, Turner saw a spiritual battle in a total eclipse, a darkened sun. God may have well been in the theophany of this eclipse. One interpretation is the eclipse represented obscuring the truth, in this case the sin of slavery created and perpetuated by whites. After this first vision, Turner withdrew from the other slaves, again signaling he was a prophet: holy, set apart, and distinctive from those around him.

Turner's second vision revealed "the knowledge of the elements, the revolution of the planets, the operation of tides, and changes of the seasons." Again, he referenced nature, tides and seasons, created by God, perhaps bringing him closer to the mind of the omniscient God. It seems God revealed some of the mysteries of the workings of the universe to Turner. With this revelation, Turner further sought to be sanctified or holy. I am guessing he once again set himself apart through prayer and meditation.[15]

In the third vision, Turner said the Holy Spirit was with him, another reference to holiness. Turner saw himself standing in the heavens or sky. Turner was in the presence of God. In the sky, he saw the Lord Jesus Christ's lighted hands outstretched as if on the cross, one hand pointed to the east and the other to the west. Turner expressed the duality of the sun rising in the east as light and the sun setting in the west representing darkness. Christ was surely the light, and the sinfulness of humankind was darkness, the latter representing the ruthlessness of slaveholders. Turner also identified with Christ dying on the cross for the forgiveness of sins. Turner was willing to die, to sacrifice himself for the greater good of the slave rebellion. In this environmental tableau, Turner paralleled Christ sacrificing himself for others with Turner doing so to free the slaves.[16]

Turner's fourth and final vision moved him closer to the insurrection. While he was working in the fields, he said,

> I discovered drops of blood on the corn as though it were dew from heaven—and I communicated it to many, both white and black, in the neighborhood—and I then found on the leaves in the woods hieroglyphic characters, and numbers, with the forms of men in different attitudes, portrayed in blood, and representing the figures I had seen before in the heavens.[17]

This last vision was complex. It is possible that blood on the corn, hieroglyphic characters of battle between slaves and slaveholders, is again symbolic of Christ's blood sacrifice on the cross. Jesus was covered in blood when Roman soldiers put thorns on his head, and nails in his hands and feet. After Christ died on the cross, another soldier cut Jesus' side, and blood and water poured out.

Turner is not explicit about the meaning of the numbers. Yet he may have been influenced by biblical numerology learned from his grandmother, and black and white preachers. Numbers are sacred in scripture. One of the numbers might have been the Holy Trinity, the three: Father, Son, and Holy Spirit having emphasized Jesus Christ, the Son in the visions. One can only speculate, knowing that numbers were significant to him in the visions.

Turner interpreted the four visions holistically, circling back to Jesus' sacrifice with "the blood of Christ" that was "shed on this earth, and had ascended to heaven for the salvation of sinners, and was now returning to earth again." Turner saw the dew – it is reasonable to conclude an African interpretation of water as purifying – on leaves and the black and white figures and spirits in the blood. Again, this is reference to Christ's blood. With these references and images, Turner believed God gave Turner permission to revolt. Taken further, the "Saviour was about to lay down the yoke he had borne for the sins of men, and the great day of judgment was at band (sic)." Being figuratively forgiven of one's sins by Christ is like literally taking off the yoke, a wooden piece across the necks of a pair of animals then connected to a plow to work the fields. That was freedom. Rebellion… sin…removing the yoke represented ending the sin of enslavement.[18]

Turner's words and actions reverberate in the present. Melanie L. Harris says eco-womanism has been framed in part by "the colonization of the earth and the colonization of black and African bodies throughout the history of the transatlantic slave trade up to the present." The same is true of black eco-theology across time. Turner is early in the continuum of enslavement as colonization, a connection of past to present through a black theology of place. The continuum moves forward today as the liberation of people of African descent continues in the 21st century in connection to the environment.[19]

What is the connection between past to present? Turner was making his way to the armory in Jerusalem in Virginia to take guns to continue the insurrection. Although he never made it to Jerusalem, Turner may have remembered ancient Israel as a contested occupied place. The Israelites firmly entrenched themselves, having fought for the land, conquered their enemies, and ultimately held dynastic rule as kings through father David and son Solomon.

Connecting to the present in the modern Gaza-West Bank in Israel, Jerusalem, a Promised Land, was and is claimed by both Israelis and Palestinians. Making a parallel, dueling claims by blacks and whites, which included freedom and the land, echo tensions between contemporary Jews and Muslims over the ancient city of Jerusalem. Turner never made it to Jerusalem, his Promised land. Blacks are still trying to make it to the Promised Land as police continue to be violent against blacks cities across the US. In 21st century reflection, there are parallels with Jews and Palestinians in conflict.

Imagine the resistance in other contemporary racialized places and times in the US echoed in Hampton, Virginia, and ancient and contemporary Jerusalem in Israel. In the 21st century, some built environments of cities and suburbs dotted across the United States are places of racial tension, discord, and violence. Imagine Nat Turner fighting to dismantle racism through Black Lives Matter in Baltimore. Imagine the anger of the black community when a boy was dehumanized: shot by police and his dead body left for hours on the hot asphalt in Ferguson, Missouri… Imagine resistance through farmers markets springing up in urban food deserts where there are no fruits and vegetables in the supermarkets for impoverished blacks...Imagine blacks demonstrating and demanding the government clean up tainted water flowing through the taps into homes in Flint, Michigan.

Looking at a specific example, Eric Garner, a black man died when a police officer held Garner in a chokehold in Staten Island in 2014. Consider the connections among violence against blacks, injustice, and place looking to Garner. The autopsy showed that Garner died from the chokehold. Harris made the important environmental leap and connection, arguing Garner repeatedly said he could not breathe attributable to asthma, which he probably developed from pollution living in Staten Island. We have a prophetic call to respond to death caused by asthma because of climate change and choking due to white on black violence against an urban backdrop or environment.[20]

In these marginalized places and among oppressed people like Garner, what are the environmental signposts and prophetic call to resist injustice? Remember Nat Turner who read the environmental signposts that led to a slave rebellion. Blacks continue to fight for freedom; and that includes life-giving fresh fruit at farmers' markets and clean water in black neighborhoods even when the people in those places are not valued.

New prophets are creating their own 21st century signposts. Secular activists in suburban and urban places are speaking up like the prophets Samuel and Nat Turner in their own times. Young people are listening to these new prophets when some organized religious institutions are mute about justice and fail their communities.

Activists in Black Lives Matter have responded to racial injustice against blacks in urban and suburban settings. Black Lives Matter has cried out in lament of civil disobedience with black communities much like the prophet Moses with the Hebrews. Alicia Garza, Parrisse Cullors, and Opal Tometi, founders of Black Lives Matter, are some of the contemporary prophets. Though secular, Black Lives Matter has spoken and have been heard with more success than some established black churches' ineffectual connections to black communities. Black Lives Matter continues in the prophetic tradition of Samuel and Nat Turner. Black Lives Matter remains a modern prophetic voice in responding to brutalized black bodies on racialized landscapes.

Black Lives Matter first discovered its call, yes a prophetic call, at one contested signpost, a suburban place called Miami Gardens where George Zimmerman shot Trayvon Martin. The movement quickly went national as Black Lives Matter and so many others demonstrated on behalf of blacks, many of them men murdered in places like Ferguson, Baltimore, and Staten Island.

The backdrop for Black Lives Matter have been communities of abandoned lots, overcrowded housing, and homelessness – markers of impoverished urban neighborhoods, racialized landscapes. These have also been places of hope...Community centers surrounded by flower and vegetable gardens...Pocket parks that provide shade on warm summer afternoons.

The places are damaged or environmentally on the upswing as people fiercely hold onto their humanity even when whites dehumanize blacks in slavery and in freedom. Environmental justice is on behalf of marginalized people, some slaughtered against the backdrop of built city streets and natural environments of lynching in the woods. Property is property, chattel is chattel in the past, present or future. Whether that chattel was Turner or Garner, their lives, their black bodies, have been debased like the earth. The injustice against black people, black bodies and the planet cannot be separated from one another.[21]

Yet some in the church speak prophetically about broken black bodies and the fragile environment. Green the Church is a black-led Christian environmentalist group leading black churches in their communities. Under the leadership of Ambrose Carroll, they train churches to spread the word about caring for God's creation; promote sustainability of resources; and advocate for change on behalf of communities by reaching out to the government focusing on climate change.[22]

The call for environmental justice by the prophets continues. Remember Shiphrah and Puah who likely escaped with the Hebrews hoping to reach the Promised Land. Remember Nat Turner who attempted to get to his Jerusalem, his Promised Land and freedom. Remember Black Lives Matter who on urban terrains sought the Promised Land of justice for blacks in the 21st century. Remember Green the Church striving to protect blacks from the impact of climate change including soaring temperatures in urban settings. Out of a black eco-theology may we vow to act as did Nat Turner, Harriet Tubman, and Green the Church.

Notes

1 It was likely Nat Turner read the King James Version of The Bible.
2 Exodus 1:15-17, NRSVUE.
3 Dianne D. Glave, "Black Environmental Liberation Theology." *To Love the Wind and the Rain: African Americans and Environmental History*, Dianne D. Glave and Mark Stoll, eds. (Pittsburgh: University of Pittsburgh Press, 2006) and Melanie L. Harris, *Ecowomanism: African American Women and Earth-Honoring Faiths* (Maryknoll, NY: Orbis Books, 2017); Dianne Glave, "The Green Confessions of Nat Turner: The Rich and Varied Roots of Black Environmental Liberation Theology," *Sojourners* (May 2016), 10–11.

Delores S. Williams, *Sisters in the Wilderness: The Challenge of Womanist God-Talk* (Maryknoll, New York, 1993), 110.

4 Herbert Aptheker, *Nat Turner's Slave Rebellion Including the 1831 "Confessions"* (New York: Dover Publications, 1996, 2006); Patrick H. Breen, *The Land Shall be Deluged in Blood: A New History of the Nat*

Turner Revolt (New York: Oxford University Press, 2015); Eric Foner, ed., *Nat Turner: Great Lives Observed* (New Jersey: Prentice-Hall, Inc., 1971); Kenneth S. Greenberg, ed., *Nat Turner: A Slave Rebellion in History and Memory* (New York: Oxford University Press, 2003); Stephen B. Oates, *The Fires of Jubilee: Nat Turner's Fierce Rebellion* (New York: Harper Perennial, 1975); and Nat Turner, "The Confessions of Nat Turner, The Leader of the Late Insurrection in Southampton, VA," http://docsouth.unc.edu/neh/turner/turner.html.

5 Turner, "Confessions."
6 Turner, "Confessions."
7 Ephesians 6:5, KJV.
8 Leviticus 18:3-4, KJV.
9 Turner, "Confessions."
10 Harris, 26.
11 John 14:2, KJV says, "In my Father's house are many mansions: if it were not so, I would have told you. I go to prepare a place for you."
12 Harris, 24 and Turner, "Confessions."
13 1 Samuel 7:3-17, NRSV.
14 Turner, "Confessions."
15 Turner, "Confessions."
16 Turner, "Confessions."
17 Turner, "Confessions."
18 Turner, "Confessions."
19 Harris, 52.
20 Harris, 77.
21 Harris, 136 and 20 and Ta-Nehisi Coates, *Between the World and Me* (New York: Spiegel & Grau, 2015), 9–10.
22 "Green the Church," http://greenthechurch.org/about-us/carroll-ministries/.

5 Flight

> In the time of King Herod, after Jesus was born in Bethlehem of Judea, wise men from the East came to Jerusalem, asking, "Where is the child who has been born king of the Jews? For we observed his star at its rising, and have come to pay him homage."
>
> –Matthew 2:1-2, NRSVUE

In the historical record, she has no name, one of many who sought the North Star. The celestial object symbolized her escape, a vivid heavenly guide for the enslaved to freedom. She lived prior to the Civil War, in a time when runaways followed the star's light from the enslavement of the South to the promised land in the North. She followed the North Star to Canada through rain and snow, cold and heat, darkness and light. She survived. She endured. Ultimately, she prospered starting her own business laundering clothes and a family in freedom.

Flight and freedom, part of which was navigated by a star, is a distinctive black eco-theology of wilderness among blacks including the fictional women who could have been Harriet Tubman. She fled from the brutality of unpaid labor. Whites enslaved blacks through violence reinforced by Christianity shaping both white and black eco-theological perspectives. As noted in the previous chapter, whites took theological license broadly interpreting the meaning of "slaves obey your masters," drawn from Ephesians 6:5. Whites benefited by forcing slaves to work without wages to cultivate crops on land owned by whites. More importantly and simultaneously, Tubman, at least one enslaved woman through her own agency, expressed a black eco-theology retaining distinctive aspects of black religious culture and spiritual experiences in nature. All of this was grounded in individual and collective relationships with God.[1]

Enslaved, Harriet Tubman espoused a black eco-theology of wilderness relying on the star to lead slaves North. She escaped and returned to Maryland pressing other slaves onward into marshes, through forests, and across rivers. She was one of many blacks looking to the North Star, the promise of freedom in places like Philadelphia. She followed the star, also

DOI: 10.4324/9780429200700-5

at times using moss on the north side of trees, as a guide to the North. Tubman believed God gave her a lighted beacon to find her way. In 1859, she failed to lead her three brothers from Maryland. She

> was now left alone, but after watching the retreating forms of her brothers, she turned her face toward the north, and fixing her eyes on the guiding star, and committing her way unto the Lord, she started again upon her long, lonely journey … And so, with only the North Star for her guide, our heroine started on the way to liberty.[2]

Josiah Henson, also enslaved, pondered a wilderness experience theology based on the North Star and the Star of Bethlehem. Both incandescent lights represented freedom. One star was tangible and the other divine, the latter representing liberation including Jesus's birth as outlined in Matthew 2:1-2:

> I knew the North Star – blessed be God for setting it in the heavens! Like the Star of Bethlehem, it announced where my salvation lay. Could I follow it through forest, stream, and field, it would guide my feet in the way of hope. I thought of it as my God-given guide to the land of promise far away beneath its light. I knew it had led thousands of my poor, hunted brethren to freedom and blessedness.

Freedom looking to stars did become reality for some including Henson and Tubman.[3]

But were the stars real? The North Star or Polaris has clearly been identified by astronomers and can be seen on clear nights with and without a telescope. As for the Star of Bethlehem, scientists hypothesize about its existence based on three competing theories: it was not a star but the Sun, Earth, Jupiter and Saturn in alignment; it was a comet hurtling through the sky; or it was the "light from the birth of a new star, or nova." Real or imagined, both stars were metaphors for freedom for the enslaved who looked up to the sky and to the North. Henson's vision of stars illuminated one's faith and the way up North for runaways. Henson and Tubman were two of those runaways.[4]

The scriptural context for the enslaved, perhaps even for Tubman, were the Magi or Three Wise Men. These men followed the Star of Bethlehem to the town by the same name to honor Jesus' birth. The Magi asked, "Where is the child who has been born king of the Jews? For we observed his star in the east and have come to pay him homage." According to scripture – Jesus was born a baby – later sacrificed himself in death as an adult, as the Messiah, as the King of Kings to save humanity from their sin. Jesus' birth under the Star of Bethlehem represented spiritual liberation for Christians including the enslaved. The latter looked for more practical escape too.[5]

Harriet Tubman's parents must have certainly shared classic biblical narratives and scripture including the Star of Bethlehem with all their children. Both mother and daughter were illiterate like so many other slaves who were barred by slaveholders from reading and writing. The reason: it was easy enough for slaves who could read and write to forge a slave pass to escape white farms and plantations to freedom. Yet the enslaved relied on oral traditions to share scripture with one another. Bradford, Tubman's first biographer says,

> Brought up by parents possessed of strong faith in God, she had never know (sic) the time, I imagine when she did not trust Him, and cling to Him, with an all-abiding confidence. She seemed ever to feel the Divine Presence near, and she talked with God 'as a man talketh with his friend.'

Out of oppressive white culture, Tubman drew on African oral traditions sharing and contemplating scripture, and further, making complex theological meaning of biblical texts as resistance through flight in the wilderness much like Hagar abandoned with her son Ishmael in the desert in the Bible.

Tubman repeatedly referred to her foundation in scripture that continued into adulthood as documented in her many biographies: she relied on her relationship with God to see her through enslavement and as a runaway; Tubman had contact with and help from Quakers as she shepherded runaways to the North; she had a prophetic gift that she believed led her away from danger on the Underground Railroad; and Tubman attended an African Methodist Episcopal Church of Zion (AMEZ) church in Auburn, New York in the latter part of her life. Her words mirrored her religious and spiritual practices: "As she recovered from this long illness, a deeper religious spirit seemed to take possession of her than she had ever experienced before." She literally "prayed without ceasing." Tubman added: "Pears like, I prayed all de time," she said, "about my work, eberywhere; I was always talking to the Lord."[6]

Tubman was prayerful in the tradition of Old Testament prophets. She was a visionary linking Christian beliefs with her understanding of nature. God led Tubman, much like Nat Turner, "by 'warnings' from Heaven." Before Tubman first escaped "she used to dream of flying over fields and towns, and rivers and mountains, looking down upon them 'like a bird,' and reaching at last a great fence, or sometimes a river, over which she would try to fly, 'but it 'peared like I wouldn't hab de strength, and jes as I was sinkin' down, dere would be ladies all drest in white ober dere, and dey would put out dere arms and pull me 'cross.'" She realized once in the North that "these very places as those she had seen in her dreams, and many of the ladies who befriended her were those she had been helped by in her vision."[7]

Prayerful and prophetic, Tubman was no less of a contemporary public theologian compared to James H. Cone or Howard Thurman. She interpreted and entwined the biblical story of the promised land and her own narrative in the context of the wilderness. Tubman expressed a black eco-theology, specifically of the promised land following the North Star: for some black eco-theology was an understanding of and relationship with nature based on one's bond with God through the lens of escape. Part of that liberation theology was that Tubman – a Christian, runaway, and enslaved – led others to collectively rebel by escaping in the tradition of the Hebrews.

Blacks, including Tubman, turned to the metaphor of the promised land as a sacred and ethereal place to survive the mental and physical oppression of the white enslavers. The Exodus narrative of enslavement, flight, and freedom – for Hebrews set against the backdrops of built and natural environments of pyramids and for blacks tobacco fields, respectively – is still potent today among blacks long marginalized through many centuries in the United States. The Hebrews escaped Egypt, wandering the desert in the wilderness, and after forty years found Canaan, the promised land. From the biblical text, blacks found immeasurable meaning and comfort in the metaphor of the promised land. Tubman led at least 300 slaves to the hope and the reality of promised land much like the Hebrews who escaped the Egyptians. She reached the Promised Land when she first escaped to Philadelphia, Pennsylvania to safety and freedom. Tubman, like the Hebrews, sought the fulfillment of a promise of freedom provided by God. Surely if she had faith she would succeed relying on markers set by God in nature to guide herself and others from enslavement in the South to freedom in the North.[8]

Tubman's words and experiences embodied a wilderness experience theology reflected in numerous trips to the North or the promised land. In *Free Black Communities and the Underground Railroad* (2014), Cheryl Janifer LaRoche reflects on the promised Land metaphor:

> Language associated with the Underground Railroad reflected the religious and spiritual character of the movement. Harriet Tubman triumphed as 'the Moses of her people'; the flight out of bondage and Black migration mirrored the exodus from the land of Egypt; the Ohio River transformed into the River Jordan; Canada represented both heaven and the Promised Land; escapees were bound to glory.

Tubman's journey took her from Maryland to Pennsylvania, New York, and Canada.[9]

Tubman pursued this promise, this freedom for herself and other enslaved blacks drawing on her understanding of *The Bible* and nature, a nascent black environmental liberation theology (BELT), part of black eco-theology. Cone describes the foundation of BELT, a black liberation theology that draws together Christian and being black and calls for an "inner spiritual strength" to survive and further conquer oppression, including the institution of slavery. Tubman fought such racial oppression combining her faith with her knowledge of the wilderness to escape oppression by whites through the lens of BELT. Nature was the pathway to liberation for her.[10]

The promise was based on many biblical covenants God made in forging the nation of Israel including enslaved Hebrews in Egypt who became free as Israelites in Canaan. Though slaves did not use covenant language, a promise from God of freedom had similar meaning. The many covenants were Noahic, Abrahamic, Mosaic, and Davidic.

In the Noahic covenant, God promised never to destroy the earth again by flood, while confirming "the eternal bond between God and his creatures." When the slaved had nothing else, they had their relationship with God. In the Abrahamic Covenant, God promised land and descendants to Abraham in Genesis 12:2-3. The institution of slavery did not allow for the fulfillment of such a promise. Yet the enslaved wanted freedom and in turn their own land and homes. Few were landowners even after slaves were freed. With the Mosaic Covenant, God made the Israelites a priority. Exodus 19:5 says, "Now, therefore, if you obey my voice and keep my covenant, you shall be my treasured possession out of all the peoples." Ultimately, God showed favor leading the enslaved Hebrews to the land of Canaan, for blacks freedom in the North. Exodus 6:4, the Davidic covenant promised the coming of Christ who would save people from sin and death. For the enslaved, when all else failed, they believed in spiritual freedom.[11]

How might blacks have interpreted and lived into the promise or covenant of freedom expected from God, when they saw themselves much like enslaved Hebrews who one day would be free Israelites? Tubman's narrative offers some insight and answers.

The enslaved embraced and invoked the many metaphors of the Hebrews fleeing oppression and later becoming Israelites seeking freedom drawing on the biblical narrative: Moses – Tubman, the Moses of her people; Egypt – enslavement of blacks in the South; the wilderness including the Jordan River – escape by blacks through challenging spiritual, emotional, and physical places; and the promised land or Canaan – the North, for Tubman, was Auburn, New York, and St. Catherines, Canada with the emphasis here on Philadelphia, Pennsylvania. Tubman's escape was bound to that of the Hebrews, ushering the enslaved through the covenant of freedom.

The promise for some began with Tubman. She rebelled relying on nature in the tradition of the Hebrews. Sarah H. Bradford, Tubman's first biographer said, "THE MOSES OF HER PEOPLE, may seem a little ambitious, considering that this Moses was a woman, and that she succeeded in piloting only three or four hundred slaves from the land of bondage to the land of freedom." God made promises to Moses and his people. Blacks expected the fulfilment of freedom by God from white oppression. Tubman, a leader, resisted relying on God and her faith in God much like Moses: through nature she fled across many natural terrains including rivers, woods, and swamps. Tubman used her knowledge of nature to navigate the landscape to Canaan or freedom.[12]

William Grimes – born 1784, sold at 10 years old to Colonel William Thornton of Culpeper County, Virginia – echoed Tubman's theology and experiences. He described his short-term escape in 1799:

> Before I came down I felt satisfied, and it did seem to me that the Lord heard my prayers, when I was a poor, wretched slave, and delivered out of the land of Egypt, and out of the house of bondage; and that it was His hand, and not my own artfulness and cunning, which had enabled me to escape.

In attempts to escape, Grimes prayerfully ran through cold temperatures in the 40s' and 50's, pine trees, snow, and a corn field. In his final and successful attempt to be free he landed in Connecticut amplifying his faith: I went up on a high mountain and prayed to the Lord to teach me my duty, that I might know whether or not I ought to go back to my master.[13]

Before Tubman escaped seeking God's promise of freedom going through the wilderness, she began her life in Maryland, the metaphor being Egypt. She was born circa 1820: slaveholders kept spotty records on the enslaved, so the date of her birth is an approximation. Her parents Harriet and Ben Ross named their baby girl Araminta (Minty). Tubman spent her early days in a cradle carved from a sweetgum tree, an object of love produced from nature. A slab of wood crafted by the hands of her father was prescient since Tubman later worked in timber with her father and escaped slavery through the woods.[14]

Egyptians enslaved and by extension oppressed the Hebrews, something blacks also experienced. In Exodus 5:6-9, the Egyptians forced the Hebrews to maintain the same quota to make bricks without the necessary straw to bind the brick. Yet God sent Moses to save the Hebrews from enslavement. Moses demanded justice from Pharaoh. Ultimately, Moses led the Hebrews out of the oppression of enslavement forced to find their own building materials without being paid wages. Though Egypt was oppressive for Hebrews, Egypt was in some ways familiar, filled with religious rituals, family, friends, and belongings. The forced labor and familiarity reverberated for blacks where they were oppressed by slaveholders. Yet the enslaved created some semblance of home and kinship based on deep familial connections and support.[15]

Yes, Tubman's Egypt was both oppressive and ironically in some ways home: Dorchester County, part of a peninsula on the Eastern Shore of Maryland in the Upper South. Catherine Clinton described the locale, a place of contradictions where Tubman's parents first lovingly rocked to sleep, and slaveholders brutalized her. The land was on:

> Rich, rolling fields semi-circling Chesapeake Bay, abutting Delaware to the east and grazing Pennsylvania to the North. Fields dappled with sun and lush with grain were crisscrossed by dozens of waterways throughout the peninsula, joining rivers flowing from marshes out to the beckoning saltwater. Waterfowl and wildlife were abundant, offering hunters as rich a harvest as that gathered by those who cultivated the land (with) oyster beds that spread underneath the water.

Clinton's softened version of Dorchester County more aptly fits the more idyllic experiences of slaveholders. Yet her description is insight into Tubman's meaning of place.[16]

Dorchester County was different, more pragmatic for the enslaved compared to Clinton's vision of place. Tubman lived on a coastal plain, which was flat, low land off the Maryland coast of the Chesapeake Bay. Marshes and waterways made for a humid place. Soil of the coastal plain ranged in quality from excellent – well drained and rich to poor – sandy and unproductive. Whites profited from timber worked by the enslaved like Tubman on Dorchester County's poorer soil. She was also surrounded by nature – seagrass, cattails, geese, and ducks--probably with little leisure time to appreciate the beauty. Still Clinton says,

> She learned to love the land, where flora and wildlife reflected seasonal change. The skunk cabbage would bloom in early spring, sometimes as early as February. Whippoorwills would serenade on summer evenings, and during autumn Canada geese might squawk overhead while migrating south.[17]

Beyond the fields, Tubman and her family lived and worked on parts of the farm. They probably planted vegetables in the gardens of the enslaved in the summer heat while swatting at stinging insects. The gardens would have been dotted with tomatoes, potatoes, cucumbers, and okra to supplement the meager rations supplied by the slaveholders. Imagine animal pens, perhaps a barn, with pigs in styes and stables with horses. Like others who were enslaved, the family probably lived in a crude slave cabin that was windowless, kept warm with fire in a hearth in the fall and winter, and unrelentingly humid and hot in the spring and summer.

In this place, slaveholders used Tubman and other enslaved as unpaid laborers to tend crops. During the colonial period from the 17th to the mid-18th centuries, slaveholders relied on this labor to profit, planting the monoculture of tobacco, a lucrative cash crop. The process was complex and labor intensive: germinating tobacco in cold beds, fertilizing the soil with manure, planting in the fields, topping off the flowers of tobacco, and harvesting the leaves to be hanged to cure for several weeks. When slaveholders exhausted the fertility of their tobacco fields, they burned trees for new fields. By the 1750s tobacco became less lucrative when prices dropped, and slaveholders experienced financial losses. Whites were forced to diminish their reliance on tobacco.[18]

White planters like the Brodess family, who owned the Tubmans, replaced labor intensive tobacco with corn that was ultimately exported to the Caribbean. Corn production was simpler compared to tobacco: slaves composted the soil with manure, later tilling the earth with a pitch fork or spade to plant corn seeds in rows; they weeded periodically; and slaves then harvested the corn by plucking ears from the stalks and later shucking the husks; and if the Brodess family owned a corn harvester then less human labor was required during harvesting meaning higher productivity and profits.[19]

The slaveholders shifted monocultures from tobacco to corn. Such change further complicated the lives of the Tubmans and others that were enslaved because family and friends were chattel sold for profit. Slaveholders like the Brodess family needed fewer slaves, a surplus of labor, as they shifted from cultivating tobacco to corn. The Brodess family sold slaves to whites down South where demand for slaves was higher with cotton and sugar cultivation and production. Tubman's brother was almost sold. Harriet, their mother, hid him away until the slaveholder gave up on the sale. Later, three sisters were also sold. Tubman saved other family members from sale leading them North. All of this was a version of defying Egypt.[20]

Tubman's difficult life might have prepared her to resist whites. As a child, Tubman cared for a white child. Tubman often rocked the baby to sleep. Unfortunately, Tubman fell asleep. The white mistress and the mother of the baby responded by beating Tubman. According to Bradford,

> It may be that she was thus prepared by the long habit of forced wakefulness, for the night watches in the woods, and in dens and caves of the earth, when pursuers were on her track, and the terrified ones were trembling in her shadow.[21]

Tubman was being prepared for her first short-term escape that would later take her further afield to the North. As a child, she attempted to steal some sugar to eat, a mistress – Tubman was on loan – attempted to beat Tubman. Rather than face a whipping, Tubman hid in a pigpen fighting the animals for scraps. The story has a biblical ring to it: the prodigal son who demanded his inheritance from his father, spent all of it, ended up feeding pigs and eyeing their food. Sadly, there was no inheritance, just the brutalization of a child who used the environment to escape the violence of whites.[22]

Much like Tubman, Linda Brent escaped slavery looking to nature and religion tied to a name. Brent wrote a slave narrative under the pseudonym of Harriet Jacobs. She lived from 1831 to 1897 and was born a slave in Edenton North Carolina. Brent devised many plans to resist the sexual advances of the man who owned her. She considered fleeing, like so many women both enslaved and sexually abused. Seeking solace, she sought out the ancestors for spiritual support, referring directly to her mother. Brent said, "When I entered the church, recollections of my mother came over me, and I felt subdued in the spirit."[23]

Brent turned to her parents and slave insurrectionist Nat Turner, for answers to remain enslaved in the South staying close to her children or running North to be free. She did all this against the backdrop of nature referring to a tree stump. The scene was outside a church:

> The graveyard was in the woods, and twilight was coming on. Nothing broke the deathlike stillness except the twitter of a bird. My spirit was overawed by the solemnity of the scene … It never seemed so sacred as now. A black stump, at the head of my mother's grave, was all that remained of a tree my father had planted … I knelt down and kissed them, and poured forth a prayer to God for guidance and support on my perilous step I was about to take.

Trees were sacred in African spirituality, a link to her African roots, adding to invoking her ancestors.[24]

Like Brent, Tubman transformed herself from someone exploited – Brent was sexually harassed by a slaveholder and Tubman beaten as a child – to an undeterred and dynamic leader. Tubman began as Araminta Ross and become Harriet Tubman, Moses of Her People. By 1844, she was in her early twenties married to John Tubman, a free black man. She kept his last name even when he deserted her to marry another woman. Perhaps her commitment to Christianity including the institution of marriage had her keep his last name; and just before her first escape in 1849 from the Brodess plantation, she changed her first name to Harriet, taking her mother's name. In doing so, Tubman was probably influenced in the African tradition of honoring and respecting African elders, tracing matrilineal ancestry and by extension venerating the ancestors.

Tubman's newly chosen names reflected a metamorphosis from enslaved to runaway, victim to rebel and as part of that transformation, she ran away from the slaveholder and unremitting brutal unpaid labor into the unknown of the wilderness. Imagine courageously abandoning familiar places and people for something new, frightening, and dangerous while still in her mid-twenties. Tubman first escaped in 1849, and later between 1852 and 1861. She returned numerous times to Dorchester County leading as many as three hundred slaves to the North.

Between Egypt and Canaan, between Dorchester County and Philadelphia, Tubman's narrative continued as she left behind Pharoah or the slaveholders. She experienced a wilderness experience theology, "a near-destruction situation in which God gives personal direction to the believer and thereby helps her make a way out of what she thought was no way."[25]

In the Old Testament, Hagar reflects the biblical wilderness experience

> in a desolate and lonely wilderness where she—pregnant, fleeing from the brutality of the slaver Sarai, and without protection—had religious experiences that helped her, and her child Ishmael survive when survival seemed doomed. For both Hagar and black women, the wilderness experience meant standing utterly alone, during serious trouble, with only God's support to rely upon.

Abraham, a Hebrew patriarch, abandoned Hagar and his son Ishmael in the wilderness of Beersheba in the Negev Desert in what is now known as southern Israel.[26]

Tubman's Beersheba was the journey from Maryland through Delaware to Pennsylvania. She risked her life in the wilderness relying on God to take her to civilization. Tubman propelled herself toward freedom. She survived like Hagar while meeting Jesus physically and metaphorically in the wilderness based on her faith in God.[27]

Tubman dreamed of the wilderness as a physical place and spiritual experience. In one instance, she decided not to participate in the 1859 John Brown Raid on Harper's Ferry in Virginia based on a dream. Tubman escaped death "a near-destruction situation in which God gives personal direction to the believer and thereby helps make a way of what she thought was no way" in the aftermath of the raid. Tubman saw John Brown, the old white man, who led a rebellion taking on "the burden of a despised race" of blacks:

> She thought she was in 'a wilderness sort of place, all full of rocks, and bushes,' when she saw a serpent raise its head among the rocks, and as it did so, it became the head of an old man with a long white beard, gazing at her, 'wishful like, jes as ef' he war gwine to speak to me,' and then two other heads rose up beside him, younger than he,—and as she stood looking at them, and wondering what they could want with her, a great crowd of men rushed in and struck down the younger heads, and then the head of the old man, still looking at her so 'wishful.'

Tubman envisioned a failed attack, finding answers in the wilderness: the death of Brown who was captured and hanged, her own escape from hanging, and the challenges of nature each time she ran away on her own and later with others.[28]

God continued to safeguard Tubman from destruction in the wilderness when she first escaped in 1849 from the Brodess plantation. She understood nature more each time she traveled through the wilderness trekking at night and resting by day to avoid capture. Tubman

> might have sought her daily rest near a hollowed-out tree, looking for a nest of brown bats, as they would gobble up the pesky mosquitos that plagued her. She would try to fade into the landscape during sunlight, perhaps refreshing herself with provisions such as dried muskrat (called marsh rabbit by locals).[29]

Tubman also experienced the wilderness drawing on the Hebrews on the run from Egypt. She relied on an

> invisible pillar of cloud by day, and of fire by night, under the guidance of which she journeyed or rested. Without knowing whom to trust, or how near the pursuers might be, she carefully felt her way, and by her native cunning, or by God given wisdom.

Bradford's language describing Tubman's experience is very much the same as Exodus 13:21–22:

> The LORD went in front of them in a pillar of cloud by day, to lead them along the way, and in a pillar of fire by night, to give them light, so that they might travel by day and by night.[30]

Tubman's route was about ninety miles – a short distance by car today but a difficult journey during enslavement – taking ten to twenty-one days from Maryland to Philadelphia. Clinton says,

> Most likely Tubman took what had become the most common route for fugitives from the region: northeast along the Choptank River, which reaches far inland, cutting a swath across the verdant Delmarva peninsula (shared by western Delaware, eastern Maryland, and a small offshore slice of Virginia to the southernmost point). She later confided that she had observed that all the streams she knew ran north to south.

Tubman might have used the direction of flowing water as a guide during her first foray.[31]

The Choptank River was Tubman's Jordan: the latter North of the Dead Sea on the path to Canaan, the Promised Land. Joshua led the Israelites on the final stage out of the wilderness through the Jordan. Each time Tubman crossed the Choptank she might have invoked the River Jordan, a code for escape for slaves based in *The Bible*. The story of crossing the River Jordan was central to the spiritual lives of many enslaved. Moses never got his chance to cross the river dying just before reaching Canaan. Instead, Joshua led the Hebrews to the land God had promised to Moses and his people. Tubman, like Joshua and unlike Moses, took her people to the other side to freedom. In Joshua 3-4, priests waded in the Jordan River, the waters were held back, and the people were able to cross. When the Israelites, formerly Hebrews were enslaved, landed safe on the other side the waters rolled again filling the gap killing the Egyptians. Bradford gives some insight to Tubman's interpretation of the Jordan so familiar in Negro spirituals the enslaved sang:

> When datare ole chariot comes, I'm gwine to lebe you, I'm boun' for de promised land, Frien's, I'm gwine to lebe you … I'll meet you in de mornin', When you reach de promised land; On de oder side of Jordan, For I'm boun' for de promised land.[32]

Crossing the Choptank in Maryland, Tubman led slaves through the heat and humidity of Maryland and Delaware, the sudden snowstorm in Maryland, and hard rain in Delaware. The weather changed depending on the season with temperatures below freezing in the 20's Fahrenheit and muggy often in the high 80's Fahrenheit. The enslaved including family and friends trudged after her through hills and valleys in the Piedmont Plateau of northern Delaware.[33]

On the journey, Tubman and the other fugitives relied on the good will of conductors of the Underground Railroad for food and shelter. Runaways tucked themselves into trees to allude white's searching for them. According to Bradford,

> They had been long without food, and were nearly famished; and as the pursuers seemed to have passed on, Harriet decided to make the attempt to reach a certain 'station of the underground railroad' well known to her; and procure food for her starving party. Under cover of the darkness, she started, leaving a cowering and trembling group in the woods, to whom a fluttering leaf, or a moving animal, were a sound of dread, bringing their hearts into their throats.

The wilderness was full of dangers. Tubman "must have dreaded the bloodhounds signaling a posse in pursuit. Would she have known to rub asafetida (a foul-smelling herb) on her feet to elude tracking dogs?"[34]

The rough terrain of the mountain passes left those who followed "foot-sore, and bleeding." The sweetgum trees proved another challenge. The seed pods were "covered with spiny, prickly burrs" that "pierce the calloused, unprotected feet" of the runaways with the children's small feet the most vulnerable. Sighting foxes, chipmunks, rabbits, squirrels, bass, and trout was easier to bear than bleeding, along with sleeping on pine needles or moss in the woods.[35]

Tubman understood and experienced freedom through the lens of a Canaan theology: objective covenantal language of place interpreted as justice. The Hebrews were first oppressed and later freed in Canaan, for Tubman Philadelphia, a new land promised by God in covenant. Through Moses God led the Hebrews "out of the misery of Egypt" leading them to Canaan, "a land flowing with milk and honey." God agreed to literally give the Hebrews land, justice after being enslaved by the Egyptians. Blacks realized the fulfilment of justice arriving in the North, comparable to Canaan. The Israelites entrenched themselves in Canaan establishing the kingdoms of Israel and Judah. Tubman's Israel was freedom from enslavement and her Judah using the North as her launching point to lead others to freedom.[36]

Tubman, like the Israelites, arrived at the boundary between wilderness and civilization crossing into Philadelphia in 1849. Later in 1851, she was prompted by her vision that represented freedom, God's promise to blacks fulfilled, to go back to Dorchester County, Maryland for her brother. In the vision, Tubman saw children were torn from women, and she heard them cry out in anguish. Tubman dreamed she was welcomed to the North:

> And beckoning hands were motioning her to come, and she seemed to see a line dividing the land of slavery from the land of freedom, and on the other side of that line she saw lovely white ladies waiting to welcome her, and to care for her. Already in her mind her people were Israelites in the land of Egypt, while far away to the north somewhere, was the land of Canaan.

Tubman revisited Egypt but it is ultimately Canaan or Philadelphia that is significant in this passage.[37]

Tubman ultimately crossed the Mason Dixon Line dividing the slave state of Maryland from the free states of Delaware and Pennsylvania. The promise of freedom was in Canaan. Vincent Harding describes a better life but not quite a land of milk and honey for blacks:

> Canaan was a strange land. In many crucial ways, it had been easier to understand black struggle in the South. At the most personal levels, the objective there was to break with slavery and move to a land where the institution was no longer legal, defying the dominance of the slaveholders over black lives.

Though Tubman was free, she faced many challenges in the northern city of Philadelphia: racism was prevalent; she was lonely having left behind familiar places and people in Dorchester County; and she knew she could be recaptured and returned to enslavement.[38]

After her arrival in Philadelphia in 1849, Tubman had to adjust to the life of a growing city with "over half a million people and claimed to be the fourth largest metropolis in the world—after London, Paris, and New York." Most prominent in the city was a busy seaport with "flourishing marine industries, manufacturing, and mercantile interests, which tied the city to a global market." She walked streets laid out with the careful urban planning of tree-lined streets and avenues on a grid of quadrants. With four very distinct seasons typical of the Northeast, the most difficult adjustment were the winters. Did Tubman have a coat? Sturdy winter boots? Protection from cold and ice was doubtful because she probably arrived with no money and based on her time in St. Catherines in Canada she and others with her did not have protective clothing. Philadelphia was on a plain between the Schuylkill and Delaware Rivers. The port accessed from the Delaware River to the Atlantic Ocean. Ultimately, Tubman's first impression of the city was positive connecting to her spirituality:

> 'I looked at my hands,' she said, 'to see if I was de same person now I was free. Dere was such a glory ober eberything, de sun came like gold trou de trees, and ober de fields, and I felt like I was in heaven.'[39]

Harriet Tubman was never canonized as a saint: she was not Catholic so would not have been formally recognized. Yet she lived a saint's life that was selfless and holy, constantly in communion with God. The benchmarks for Catholic sainthood is heroism – for example selflessly devoting one's life to the care of others; beatification – the church deems the deceased blissfully in heaven; and canonization individual honored as a saint. No pope, priest or pastor, no church, or even AME Zion ever conferred sainthood on Harriet Tubman. Yet history and her faith showed her to be a saint or otherwise highly esteemed here on earth, having courageously and repeatedly led enslaved friends and family North to freedom. Tubman was the Moses of her people who lived a Promised Land black eco-theology that should be recognized today as a saint or otherwise.[40]

Still some little black girls in the modern world can look up to Tubman as the girls can enter natural cathedrals like lakes, rivers, and forests. These places were once contested where slave trackers violently apprehended slaves. No more. The slaves fought back and resisted and so do the girls entering these sacred spaces in memory of Tubman. These little girls along with so many other blacks travel the wilderness of environmental racism with the hope of a Canaan or justice.

Notes

1 Ephesians 6:5, NRSVUE.
2 Kate Clifford Larson, *Bound for the Promised Land: Harriet Tubman, Portrait of an American Hero* (New York: Ballantine Books, 2004), 78–79 and Sarah H. Bradford, *Harriet: The Moses of Her People* (New York: Geo R. Lockwood & Son, 1886), 29, Electronic Edition, "Documenting the American South," Library of Congress, https://docsouth.unc.edu/neh/harriet.harriet.html.
3 Matthew 2:1-2, NRSVUE; Cheryl Janifer LaRoche, *Free Black Communities and the Underground Railroad* (Urbana: University of Illinois Press, 2014), 88–89; and Josiah Henson, *Autobiography of Josiah Henson: An Inspiration for Harriet Beecher Stowe's Uncle Tom* (Mineola, NY: Dover Publications, 2003), 78.
4 Victor Gill, "Star of Bethlehem: The Astronomical Explanations," *BBC News Magazine* (December 23, 2012), https://www.bbc.com/news/magazine-20730828 and Geza Vermes, *The Nativity: History & Legend* (New York: Doubleday, 2006), 94–96.
5 Hebrews 9:12, NRSVUE; Matthew 2:2 (NRSVUE); and Matthew 2:6 (NRSVUE).
6 Catherine Clinton, *Harriet Tubman: The Road to Freedom* (New York: Little, Brown and Company, 2004), 20 and 92 and Bradford, 24.
7 Bradford, 114–115, 114, and 115.
8 Clinton, 37; George Sullivan, *Harriet Tubman* (New York: Scholastic, Inc., 2001), 43–44; and M. W. Taylor, *Harriet Tubman: Antislavery Activist* (New York: Chelsea House Publishers, 1991), 37.
9 LaRoche, 88.
10 James H. Cone, *A Black Theology of Liberation* (Maryknoll, NY: Orbis Books, 1990, 2020); James H. Cone, *The Cross and the Lynching Tree* (Maryknoll, NY: Orbis Books, 2013); Gustavo Gutierrez, *A Theology of Liberation: History, Politics, and Salvation* (Maryknoll, NY: Orbis Books, 1988); and Dwight N. Hopkins, *Introducing Black Theology of Liberation* (Maryknoll, NY: Orbis Books, 2004).
Cone, xvii and xviii.
11 Harold W. Attridge, ed., *The HarperCollins Study Bible* (San Francisco: HarperSanFrancisco, 1989), 17; Genesis 12:2-3, NRSVUE; Exodus 19:5, NRSVUE; and Exodus 6:4, NRSVUE.
12 Bradford, 3.
13 William L. Andrews and Regina E. Mason, *Life of William Grimes, the Runaway Slave* (Oxford: Oxford University Press, 2008), 57 and 43.
14 Michael E. Ruane, "Harriet Tubman Fled a Life of Slavery in Maryland." *The Washington Post* (March 4, 2017), https://www.washingtonpost.com/local/harriet-tubman-fled-a-life-of-slavery-in-maryland-now-a-new-visitor-center-opens-on-the-land-she-escaped/2017/03/03/1134955e-fd0e-11e6-8ebe-6e0dbe4f2bca_story.html?noredirect=on&utm_term=.97c350e0e287.
15 Exodus 5:6-9, NRSVUE.
16 Clinton, 3.

17 James DiLisio, *Maryland Geography: An Introduction* (Baltimore: John Hopkins University Press, 2014), 59 and Clinton, 20.
18 DiLisio, 21.
19 Clinton, 3.
20 Clinton, 94.
21 Bradford, 21.
22 Luke 15:11-32, NRSVUE.
23 Harriet Jacobs, *Incidents in the Life of a Slave Girl* (New York: Signets Classics, 2010), 101 and 87.
24 Jacobs, 101.
25 Delores S. Williams, *Sisters in the Wilderness: The Challenges of Womanist God-Talk*(Maryknoll, New York: Orbis Books, 1993), 108.
26 Williams, 109 and Genesis 21:14, NRSVUE.
27 Williams, 109 and 113.
28 Williams, 108 and Bradford, 118.
29 Clinton, 37 and Bradford, 29.
30 Bradford, 30 and Exodus 13:21-22, NRSVUE.
31 Clinton, 38 and 37.
32 Attridge, 310; Clinton, 38, 37, 84; Joshua 1:2-6, NRSVUE; and Bradford, 27 and 26.
33 Katherine B. Moose, *Uniquely Delaware* (Chicago: Heinemann Library, 2004), 7.
34 Bradford, 35 and Clinton, 34.
35 Bradford, 29–33 and Larson, xiii.
36 Exodus 3:17, NRSVUE.
37 Bradford, 26 and Clinton, 84.
38 Vincent Harding, *There is a River: The Black Struggle for Freedom in America* (New York: Harcourt Brace Jovanovich Publishers, 1981), 119 and Bradford, 31.
39 Bradford, 26 and Catherine Clinton, 49; "Bird's Eye View of Philadelphia," Library of Congress Geography and Map Division Washington, D.C. 20540-4650 USA dcu, 1857, https://www.loc.gov/item/75696517; Steven Conn, *Metropolitan Philadelphia: Living with the Presence of the Past* (Philadelphia: University of Philadelphia Press, 2006), 30–31 and 160; and Bradford, 26 and 30–31.
40 Sarah Gallick, *The Big Book of Women Saints* (New York: HarperSanFrancisco, 2007), 1–3.

6 Trees

> And the LORD God commanded the man, "You may freely eat of every tree of the garden, but of the tree of the knowledge of good and evil you shall not eat, for in the day that you eat of it you shall die."
>
> –Genesis 2:16-17, NRSVUE

Wangari Maathi is still alive at 77 and sitting under a sacred fig tree in her small village in Kenya. She holds her Bible reading Genesis 3, reflecting on black Adam and Eve's last hours in the Garden of Eden, the cradle of civilization, Africa. Maathai imagines Satan disguised as a serpent tempting the black couple with the fruit from the Tree of Knowledge of Good and Evil. Adam and Eve sink their teeth into the forbidden fruit though God admonished them to stay away or die.

The couple's disobedience exposed them to the knowledge of good and evil that revealed their sin, created separation from God, shamed them because of their nakedness, and banished them from the perfection of God and the Garden of Eden. Genesis 3:24 (NRSVUE) says, God "drove out the humans, and at the east of the garden of Eden he placed the cherubim and a sword flaming and turning to guard the way to the tree of life." Ultimately, God condemned black Adam to a life of hard work in the fields and black Eve to the pain of childbirth. Just as significant, God's presence receded physically from the couple – God had once walked with them in companionship in the garden – along with the couple's spiritually dying. Rather than eternal life in the Garden of Eden, they struggled outside the garden, where ultimately their bodies aged and succumbed to disease that killed them. Maathai walks away from the fig tree still reflecting on the sacredness of and the tragedy in the Garden where black Adam and Eve still live.

Shifting from the Afrofuturism or alternate black history of this Garden of Eden tableau is entry into Maathai's theological reflections. She argues that we live between the perfection of the Garden and Jesus' second coming. Maathai says to fully live in the imperfect Garden of Eden now between these two phases of perfection. This is apropos because her theology was people first, responding to the needs of the living who suffer because of climate change, the latter the secondary concern. Maathai says,

DOI: 10.4324/9780429200700-6

> Even Revelation itself describes the passing away of 'first earth' and the 'new heaven and earth' that will take its place (Rev. 21:1) verdant scene: 'Then the angel showed me the river of the water of life, bright as crystal, flowing from the throne of God and of the Lamb through the middle of the street of the city. On either side of the river is the tree of life with its twelve kinds of fruit, producing its fruit each month; and the leaves of the tree are for the healing of the nations. (Rev. 22:1-2)'[1]

Christians anticipate the return of Jesus Christ, the second coming, the second earth. The first, the perfect Eden was not destroyed and then replaced by the perfection of the second coming. Maathai says,

> I see them depicting an alternative to the gradation of the environment that has turned waters of life here on earth that were 'bright as crystal' into mud and silt, and the 'tree of life with its twelve kinds of fruit' into stumps and charcoal, and has led nations not being healed, but rather fighting one another for access to the remaining clean water and food supplies.[2]

This slow death of nature was not the Maathai way: she lived, fought, and served in the here and now, the many precious moments nurturing people and the environment using activism for the long term. She evokes a middle place, an in-between world in which we must function, survive, and excel in the Garden:

> The planting of a tree is also a reflection of, and it is hoped a stimulant toward, a new consciousness of the nature of the salvation that Christ offered. To make the connection more explicit, Christian theology draws a link between the tree in the Garden of Eden, cause of the Fall, and the death upon another tree—the cross—of the second Adam, who redeems the original sin and reverses Adam's curse.[3]

Maathai lived in real time helping to cultivate the world's garden by planting trees for the sake of people. The Green Belt Movement (GBM) was a culmination of Maathai's early influences and life's work. In 1977, she founded and launched the organization to support rural women from Kenya, though children and men benefit. Today, others continue the work with Maathai's environmental non-profit organization. She says,

> When I told the rural women about the benefits of trees—that they'd help stem soil erosion and improve health of the soil, making it possible to grow healthy foods; that they'd provide firewood, fodder for livestock, and shade; that they'd help regulate rainfall and provide a habitat for small animals and birds.[4]

GBM helped the women strengthen and conserve the environment, while giving them the means to improve their lives and of those around them. The organization offered an opportunity to the women to plant trees that helped to diminish soil erosion, provide a new source for kindling, a small amount of money for their efforts, and a way to retain rainwater in the soil. Approximately thirty million trees and counting have been planted and most remain standing in Kenya. Millions more have been planted across Africa. The trees drew on and combined precious water (H20) and carbon dioxide (CO2) to return oxygen (O2) to the air again giving life to creatures on earth. Humans, including Africans, require oxygen to breathe, breakdown food, and stay alive.[5]

Maathai was born to this environmental legacy on April 1, 1940, to her mother Wanjiru Kibicho and father Muta Njugi in Kenya. She was the third child with two older brothers Nederitu and Kibicho and two younger sisters Muringi and Wachatha. The family was Kikuyu, the largest ethnic group in her country. The men in her family followed many generations of black Kenyan farmers, one of many reasons why Maathai was drawn to the environment.[6]

The environment, including the geography, informed her life's purpose: the global movement she led to plant trees to continue to help save the planet. Maathai's birthplace was the continent of Africa, more specifically central Kenya. The country and nation of Kenya is bounded on the North – South Sudan and Ethiopia; South – Somalia; West – Uganda and Tanzania; and East – Somalia and the Indian Ocean. Kenya has diverse geography of mountains, deserts, and plains, along with the swamps of mangrove trees between the land and the ocean. Three spectacular geographic regions stand out: Mount Kenya, the highest altitude in the country; the Great Rift Valley, a trench edged with cliffs, mountains, and active volcanoes; and Lake Nakuru where yearly flamingos by the thousand come to breed. Other animal life adds to the diversity: hippopotami, lions, zebras, elephants, and giraffes.

On the continent of Africa and the country of Kenya, Maathai lived in Ihithe, a small rural village, not far from Nairobi, the capital of Kenya. The city of Nyeri was even closer. Nairobi means cool water in the Maasai language, a reference to the Nairobi River that cuts through the city.

Maathai's smaller world of her youth was for a time the closest thing to a Garden of Eden. The land surrounding the village Ihithe was verdant. She described the village as "still lush, green, and fertile." A single day could range from cold to hot. The region had consistent seasons. Fall in mid-March was marked with Monsoon rains. Winter in July was densely foggy and so "cold in the morning that the grass would be a silvery-white with frost." Not far from her village, she spent time in Aberdare National Park. Maathai was surrounded by all that was idyllic in nature.

Looking back on her childhood, Maathai said, "I came to identify trees as the answer to the environmental problems the Kenyan women faced, partly because I grew up in the countryside surrounded by trees and green vegetation."[7]

In 1920, the British began to colonize Kenya with environmental implications for Kenyans including the Matas–Maathai's family name. For blacks, the country shifted from sole ownership of their own farms to working as wage laborers for whites. Several white individuals and groups exploited black laborers monopolizing fertile farmland in the white Central Highlands in Kenya. Whites pushed out or forced indigenous blacks to carve out farms to systematically colonize and exploit the land and black migrant workers paid low wages.

The irony of environmental exploitation by whites was that the Central Highlands was a sacred place for the Kikuyu, Maathai's ethnic group in Kenya. She says,

> The undulating landscape of the Central Highlands of Kenya provided many places for Kikuyu priests to escape and commune with God or to receive inspiration …These grottos might be near waterfalls or rivers; they would be naturally covered by vegetation and thus it would be easy to secret oneself away.[8]

The roots of climate oppression and climate injustice in Kenya continued with structural imperialism and colonization by the British. In the early 1940's, whites manipulated black Kenyan men into migrating to the colonized Central Highlands farms for work. Men like Maathai's father was forced to leave their ancestral homes because the British manipulated and transformed the Kenyan economy. Whites transitioned Kenya from bartering goods among black Kenyans to a monetary system that required blacks pay taxes to the British. About 150,000 black men were part of the first wave of migration to the city and large farms. In doing so, they worked to pay the British the new oppressive taxes.

Many black Kenyan men left their families behind sending money home periodically visiting home. Whites exploited the Kenyan men, the latter receiving low wages. Whites became financially successful having blacks grow and whites sell cash crops including coffee and tea. Black Kenyans knew hands-on the difficulties of cultivating coffee, weeding the plants, spraying pesticide, and picking the beans for packaging and transportation. The pesticides surely poisoned men, particularly black Kenyans working directly with coffee plants. The transition from subsistence farming to a nascent agri-business of cash crops contributed to a downward cycle of poverty. Blacks responded through deforestation, cutting down too many trees primarily for fuel to keep warm in their homes and cook food to put on their tables. They merely tried to survive colonial and post-colonial economies and landscapes, ultimately and unwittingly contributing to nascent climate change.

Western Christianity, another method of colonization by whites, dramatically influenced black Kenyans. This made sense because embracing Christianity meant more privilege for blacks among whites. The same was true for Maathai. Both of her parents became Christians as did she. Christianity also influenced Maathai through much of her education and her environmental activism.

In 1947, Maathai's father sent his wife away so the children could go to school in their home village of Ilithe. Maathai and her brothers attended a school founded by Presbyterians with African teachers. In 1951 at age 11, Maathai moved to Nyeri and attended St. Cecilia's Intermediate Primary School run by Catholic nuns from Italy. She later attended the Loreto High School outside of Nairobi. Western Christianity deepened in Maathai's life when she attended Mount St. Scholastica College in Atchison, Kansas, also run by nuns.

In 1963, Maathai transitioned into secular education completing a Bachelor of Science in biology minoring in chemistry and German. At the University of Pittsburgh, she graduated with her master's in science in 1964. Maathai returned home finishing her last degree at what is now known as the University of Nairobi. The university hired her as a professor of zoology. After completing her education, she married Mwungi Maathai – probably in a Christian wedding and raised Wanjira, Waweru, and Muta, their children.

Christianity did influence Maathai thinking eco-theologically in part. Still though Maathai's training was in religious and predominantly white settings, she says "I was lucky to possess a mental attitude that was not constrained by judgment or religious dogmas." She also retained African influences, specifically the influences of her Kikiyu ethnic heritage.[9]

Again Christianity, a colonizing tool, was historically used to deepen power over Kenyans like Maathai. From 1952 to 1950, The Mau Mau Rebellion was a backlash against colonialism by black Kenyans against whites supported by the church. The uprising, a political and military movement by blacks against the colonizers, deeply impacted the country. In 1963, the rebellion was a catalyst for Kenyan independence from the British. The Mau Mau resistance and Kenyan independence probably influenced Maathai's movement to plant trees as a form of protest against the destruction of the planet by whites who controlled and used most natural resources and were primarily responsible for climate change and destroying nature.

Maathai resisted in her own way against white colonizers of the earth and people, black Kenyan patriarchy, and by extension government bureaucracy. She served her people first, then the natural realms of the planet. Maathai agitated against the Kenyan government who passively or actively deepened climate change contributing to deforestation, desertification, water run-off, water shortages, and soil erosion in Kenya.

Beginning locally at her home in Kenya, Maathai answered a spiritual and practical call to tend the environment, complementing her resistance against institutional power. Ihithe, her home village, transformed from an idyllic Garden-of-Eden-like place to a deforested and drought-ridden place. As an adult, Maathai experienced and witnessed climate change and vowed to reverse the destruction in Kenya and her village through the GBM. For this work, Maathai was the first African woman to win the 2004 Nobel Peace Prize, a public and international recognition. She died on September 25, 2011, and proudly an environmentalist steeped in spirituality and the love to serve until she was 71 years old.

The GBM was Maathai's life's mission, genius, and service grounded in her experiences growing up, her education, broader history, and climate change. Her spiritual and religious perspectives also influenced her efforts to save nature. A School of Wangari Maathai theology can be drawn from her publications. This is a new hermeneutic drawn from her words in *Replenishing the Earth: Spiritual Values for Healing Ourselves and the World* and *Unbowed: A Memoir*, both books environmental texts. I would argue that these are also sacred religious texts, a hybrid of African and by extension Kikuyu spirituality with her Christian beliefs. I draw together the beginnings of a systematic theology that is both general and specific to Maathai's love of nature, trees, and people.

Maathai drew from common beliefs blending African spirituality and Christianity: nature woven into all aspects of African life and the Source being both God and nature. She experienced the Source in two ways. First, she referred to the higher being as the Source from an African rather than a Christian perspective. Second, Maathai added,

> I can also point to the Source as the wellspring for all of the ideas that came. Sometimes the inspiration to act arrives as a spark; sometimes it takes the form of a process. Whether one is drawn into an action through a sudden rush of inspiration, or through the slow dawning of a realization that something needs to change, I would argue that all comes from the Source.[10]

Maathai said, "We also see God or the Source in elements of the natural world, lest we be accused of pantheism." To be clear, nature is not God and God is not nature. Maathai adds, "I never differentiated between activities that might be called 'spiritual' and those that might be 'secular.'" Historically, an African world view has been that nature cannot be separated from all aspects of life: nature/religion; nature/art; nature/economics; nature/work, and so on. Maathai probably believed and lived into this view of nature as it has been a common belief across communities in

Africa. She believed that you had to be in a certain place spiritually "to take advantage of the awakening" through the Source as God and nature.[11]

The meaning of nature through an African lens including her relationship to the Source informs a practical eco-theology: Maathai helped marginalized people in the tradition of liberation theology. Blacks and other disenfranchised people continue to liberate themselves and others from systemic oppression by whites and institutional racism. Maathai did so by organizing communities locally, nationally, and globally by planting trees to help save the planet and improve the circumstances of black Kenyans.

More specifically, she embraced and lived a Black Environmental Liberation Theology (BELT), that draws on black liberation theology, a way of thinking theologically about planting trees and the people who rely on the trees. BELT strives to eliminate environmental racism for just equitable access to natural resources and green places in the context of scripture, theology, and action. The theology is drawn from James H. Cone's article "Whose Earth is it Anyway?" groundbreaking reflections of environmental justice for the marginalized – with little or no access to profit from natural resources and outdoor amenities. Conversely, those who are privileged – who have easy access to green spaces including national parks and natural resources like water and natural gas – benefit. Maathai pushed on the intersection of colonialism, racism, and sexism practicing BELT to improve the lives of impoverished Kenyan women through tree planting.[12]

Another aspect of the Wangari Maathai School of Theology is from her Christian beliefs. Her response to the global climate crisis mirrors Christ's ministry, crucifixion, and resurrection. Considered another way, both climate change and Jesus' life are a circle of life, death, and life. Jesus' fragile earthly body started as biological life with a spirit, culminated in physical human death, and continued into a fully spiritual life. He performed multiple miracles on earth, sacrificed himself for humanity in death, made the promise of everlasting life, and is the living Word of God. To add another fold or layer to this, the Garden of Eden, much like Jesus, was perfect. In our current environmental crisis, our failing dying ecosystems fall short of Jesus' perfection, Her/His promise of life and the quintessential perfection of the Garden, with the destruction of the earth. Still, Maathai believed we are to live fully into the damaged Eden, doing our best to care for what we have now.

From life to death, from judgment to crucifixion, Christ was nailed to the wooden cross with two pieces fashioned into a t-shape. In the ancient Greco-Roman world, Jesus lived from circa 3 BC to CE 33 in Israel. Romans, those who ruled much of the Mediterranean and more including Israel, killed runaway slaves and common criminals in many ways including crucifixion on crosses. Pontius Pilate, the governor of Judea, condemned

Jesus's fate of death by crucifixion. The wooden cross was hewn from a living tree, perhaps a cedar, used for Jesus' gruesome death.

Christ suffering on the cross is a metaphor for 21st century deforestation around the world including Maathai's beloved Kenya. The death and resurrection of Christ is reflected in recurring death to life to death cycles of ecosystems, now globally in a dying stage. The death of millions of trees – in forests like the Amazon and Sequoias in California – because of climate change is the "crucifixion" of earth by humans mirroring Christ's death. One representation of our damaged waning ecosystems is what was probably a living cedar tree that was cut down and used to crucify Christ. Today, the planet is in jeopardy as was Jesus, both ultimately being killed. People continue to destroy, to crucify trees. The cycle worsens climate change because there are fewer trees to diminish destructive carbon dioxide in the ozone. The results are more air and water pollution, more floods, and extreme heat; and the cycle continues.[13]

Cone expands on Christ on the cross highlighting and connecting to the lynching tree used to brutally dehumanize and murder blacks in the United States as a spectator sport. Much of these lynchings were literally hangings by rope. Yet whites found other countless ways to murder blacks with guns, knives, and fire – all under the umbrella of lynching. Many of the murders were from the 1890s to the 1960s, post-Civil War and Reconstruction to the Civil Rights Movement. The lynching tree represents hateful racist hegemony by whites much like the crowd that condemned Jesus to death. In 1939, Billie Holiday recorded the words to Strange Fruit, a metaphor for lynching by ropes from trees:

> Southern trees bear strange
> Blood on the leaves and blood at the root
> Black bodies swinging in the southern breeze
> Strange fruit hanging from the poplar trees

This was a perversion through murders and misusing trees. Lynching from trees abated against blacks with 1960s Civil Rights. Still, whites continue to murder blacks in the late twentieth and twenty-first centuries. Some had hope in 2020 in Minneapolis, Minnesota with protests against the "lynching" of George Floyd when a white police officer pressed his knee on Floyd's neck until Floyd died. Such hope did not last long. Sadly, the lynching tree continues even metaphorically.[14]

The injustice of the environmental crisis mirrors similar injustices of Jesus' death and lynching blacks using trees in the United States. Similarly, marginalized people like those in the African Diaspora including Kenya suffer and die because of climate change. The lynching of climate change are fruits of racism. White colonizers in Africa who have contributed to climate change are responsible for low quality of life and shortened life spans of black Kenyan farmers. As noted earlier, white imperialists in Kenya transformed and controlled farming: initially black men worked their own land relying on self-sufficiency and bartering, later forced to be wage laborers in fields that belonged to whites. Whites impoverished blacks with low wages as farm workers. Because of low wages, blacks desperate for wood accelerated cutting down trees for wood to cook with and to heat their homes. The result was deforestation.

Jesus was resurrected from the dead after the crucifixion. Blacks and trees have experienced similar shortened and brutal deaths. Yet renewal means life, hope, and forgiveness for Christians. People of African descent long clung to life and hope in an inequitable world. That said, what does forgiveness mean for people of African descent when whites commit atrocities in the name of colonialism, imperialism, and racism, and taken further through environmental racism? Sadly, blacks are often too quick to forgive without a blueprint for radical change to stop white on black murder and environmental inequities. Blacks forgiving quickly and easily must remain a placeholder because more thought and discernment is required by individuals and groups. Concrete measures to avert continued destruction of the planet and people driven and worsened by the most affluent countries like the United States must first be taken before there is forgiveness.

Perhaps we can be forgiven of the sin of continuing to destroy the Garden of Eden during the 21st century with global warming. Maathai's practical means of hope was one tree at a time, striving for Edenic perfection. That life includes the liberation and restoration of the environment, pulling the planet back from the global destruction of the climate crisis. The cycle of life and death is the resurrection and life of Jesus Christ. Through a theological lens, Jesus' life and resurrection is a path to save the planet and the lives of people by planting living trees, earth's lungs. Marginalized people can benefit, more specifically stay alive and prosper with a reversal of climate change. The miracle of Jesus' resurrection extends to the hope of saving blacks from all sorts of violence including dying with the planet.

These broader strokes of the cross, the lynching tree, and BELT is context for the theology of trees. Maathai, a spiritual environmentalist, provided as service to nature and people is the hope of a garden that strives to be the Garden of Eden filled with life-giving trees. She and her service are at the foot of the wooden cross where Jesus was crucified for humanity's sins, and at the center of environmental justice grounded in a history of racism against people of the African Diaspora. Adam and Eve – the world's, Africa's and Wangari Maathai's Christian ancestors – left a terrible legacy causing The Fall, humanity succumbing to the first sin of eating the forbidden fruit; and this sin revolved in part around a tree: the Tree of Knowledge of Good and Evil. Out of this sin, can humanity be redeemed and by extension the earth maintained – Maathai's argues working what we have in this damaged world created by damaged people – by planting trees attempting to return to the black Garden of Eden?

One answer to this question is Maathai saying, the Aka, nomadic pygmies in the Central African Republic and the Republic of the Congo "may still be practicing the traditions of their ancestors, but may not be able to help explain that world to the seeker, who is trying to understand and recapture the world we *all* once lived in." She did her part in post-Eden, environment in Kenya and around the world with a theology, with a plan, with action.[15]

The template of tree theology is based broadly on Jesus meeting the practical and spiritual needs of many people conquered by the Romans. The plight of Kenyan women and so many other marginalized people parallel people's experiences in ancient sacred scripture and Jesus' ministry. Why accept a religion if one is starving or unhoused? Give people seeds for wheat and a cow for milk to live independently, rather than the wheat or cream that keeps Africans permanently reliant on charity and humanitarian aid. Plant a tree so people make their own way with fuel to cook food and warm a home.

Maathai's practical theology of trees underlies GBM's service. The movement's goal has been reforesting through the biodiversity of indigenous trees. In turn, the organization removed non-native trees. This was Maathai striving for a post-Eden in Kenya. She says,

> When I explained how important it was not to deforest Kenya's main water catchment areas and not to plant the wrong type of tree along the riverbanks, in forest, or in wetlands, it was self-evident that a severe problem—water and soil erosion—was being addressed.[16]

Native trees are more resilient against climate change and pests than non-native trees. In addition, indigenous trees – unlike the eucalyptus banned by GBM – require less water to be sustained and thrive. Maathai said,

> When I visit many parts of Kenya, for example, I see a land that has been abused and the soil degraded. Few indigenous trees or crops remain; instead, non-native plants dominate, and cash crops such as tea and coffee bushes stretch up and down hills and mountainsides.

As a result, GBM focused on reforestation of the damaged Garden, with indigenous trees in Kenya: baobab, Kenyan vitex, African teak, sycamore fig, silky oak, wormy albizia, oleanders, large fruited croton, and guarri. The acacia is both the national and indigenous tree. The trees sustain human life providing firewood, improve air quality taking in carbon dioxide and providing oxygen, conserve water by reducing runoff, lower and regulate temperatures, shelter biodiversity, provide herbal medicine, produce highly nutritious fruit, provide shade, and diminish soil erosion. The trees also are part of eco-systems helping to sustain animals, other plants, and insects.[17]

Looking to sub-Saharan Africa including Kenya, the Baobab tree connects nature, spirituality, and daily life. The tree dates to prehistoric times. This holistic African way of life is reflected in a tree that symbolizes human endurance. It is also said gods planted the trees upside down and alternately an African god tossed the tree from paradise with it landing upside down: the branches look like gnarled roots. In another example, Kikuyu, again Maathai's ethnic group, will not cut down fig trees because doing so means bad luck that will end in disaster.

The Kikuyu viewed trees in pragmatic and incorporeal ways. From a hands-on perspective, they practiced agro-forestry "that retained water and topsoil. Each tree that was left standing was called in Kikuyu 'one that resists the cutting of the forest.'" In addition,

> It needs to be reiterated that honoring and sacramentalizing trees in these communities didn't mean that they could not be cut down or utilized for ordinary purposes. In fact, they always have been, even to make sacred spaces. The temple built by Solomon is marked from oak, palm, date, and willow trees, among other natural elements.

Alternately from a sacred perspective, the Kikuyu believed spirits lived in all the trees that were cut down, with the spirits jumping to dead to a living tree. Maathai says,

> This was achieved by placing a stick against the tree to be cut down and then moving it to one that was to remain standing, or by planting another tree immediately in the same place as the felled one. Clearly, such restrictions stopped wholesale deforestation from taking place.[18]

Kikuyu priests also venerated and practiced sacrifices at the site of some fig trees. The rituals transformed the trees and its surroundings into sacred places. Many generations of Kikuyu revered trees like the priests. Maathai said, "One was obliged to remove one's sandal if you approached a tree during a ceremony or were climbing Mount Kenya." Maathai's mother told her "very clearly when I was a child that I was never got collect twigs for firewood from around the fig tree near our homestead since, she said, it was 'a tree of God.'"[19]

Maathai also drew from Christian perspectives often referencing and thinking theologically about The Bible. She notes that throughout the Torah, that the Israelites were tasked with following the "laws of the land that God has established, the land will be fruitful; if they are disobedient and take more than God allows them, the land will produce the equivalent of thorns." Maathai added "Indeed the Bible and many scriptures are replete with instructions about how to maintain the right relationship to fold, the land, animals, and one another, as determined by God's word."[20]

Maathai highlighted the dove returning to Noah's ark with a leaf when the flood subsided; Joshua honored God by placing a pillar under a tree; Abraham pitched his tents in a sacred grove to connect with God; and Ezekiel envisioned God as a tree who metaphorically produces water as a sign of God providing for God's people.[21]

From Christian and African perspective, Maathai understood theology and called people to action. Through a practical theology, she spoke to an unnamed church sometime on a Sunday in November. She said,

> In Kenya, the annual Easter holiday lasts four days, from Good Friday through to the Holy Monday. "Can't you plant one tree during those four days?" I asked the congregation, "How many of you will plant next Easter Monday?" I persisted. Everyone put up a hand. "How many of you will recall do it?" I continued. People laughed. I laughed too, as I delivered my final line. "I should be a bishop," I told them. "If I had a purple miter, you'd take me seriously." When I finished, the bishop asked the congregation another question. "How many of you feel you have been preached to today?" And they all raised their hands.

This moment was very Christian in the context of Holy Week that included Palm Sunday and Easter. Maathai added,

> The planting of a new tree—a symbolic replacing of Christ's tree of death with a tree that can bring life and food to many creatures (including humans)—offers a parallel to the Christian message of life conquering death and the older traditions of celebrating the seasonal cycle through an acknowledgement of the transition from the darkness of winter to new life.

It was also very African bringing together religion, community, and the environment: all aspects of life are intertwined and not separated from African perspectives. The meeting speaks to the resurrection of Christ coming back to life and of Africa by planting indigenous trees.[22]

Maathai words continue to give us hope for a future that binds religion and nature to one another for uplifting humanity and saving the planet:

> These values are not contained only within certain religious traditions. Neither does one have to profess a faith in a divine being to live by them. However, they do seem to be part of human nature, and I'm convinced that we are better people because we hold them, and that humankind is better people because we hold them.[23]

Notes

1 Wangari Maathai, *Replenishing the Earth: Spiritual Values for Healing Ourselves and the World* (New York: Doubleday, 2010), 125.
2 Maathai, 125.
3 Maathai, 134.
4 Maathai, 32 and "Who We Are," The Greenbelt Movement, https://www.greenbeltmovement.org/who-we-are.
5 Maathai, 32.
6 Wangari Maathai, *Unbowed: A Memoir* (New York: First Anchor Books, 2007), 3.
7 Maathai, *Unbowed*, 3 and Maathai, *Replenishing the Earth*, 31.
8 Maathai, *Replenishing the Earth*, 81.
9 Maathai, 30.
10 Maathai, 16 and 134.
11 Maathai, 16 and 31.
12 James H. Cone, "Whose Earth Is It Anyway?" *CrossCurrents*, vol. 50, no. 1/2, (Spring/Summer 2000), 36–46 and Dianne D. Glave, "Black Environmental Liberation Theology," *"To Love the Wind and the Rain:" African Americans and Environmental History*, Dianne D. Glave and Mark Stoll, eds. (Pittsburgh: University of Pittsburgh Press, 2006), 189–199.
13 James H. Cone, *The Cross and the Lynching Tree* (Maryknoll, NY: Orbis Books, 2011), xv–xvi.
14 David Margolic, *Strange Fruit: A Biography of a Song* (New York: Ecco Press, 2001).
15 Maathai, 49.
16 Maathai, 33–34.
17 Maathai, 64.
18 Maathai, 86 and 78.
19 Maathai, 78 and 79.
20 Maathai, 45.
21 Maathai, 81.
22 Maathai, 132–133.
23 Maathai, 16.

Index

Pages followed by "n" refer to notes.

Made in the USA
Las Vegas, NV
19 September 2026

fb8fe5b6-a46d-4edb-b256-ee9073a79d6dR01